The Fundamentals of Vitalism

by Prof. Gregoriy Shifrin

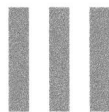

Clink Street

Published by Clink Street Publishing 2021

Copyright © 2021

First edition.

ISBN:
978-1-913340-46-9 - paperback
978-1-913340-47-6 - ebook

Contents

Vitalism is new conception of maintaining and restoration of human life forces in accordance with changing genome programs. Its methods of vital resistibility will expand the competence of physicians and help apply unique personalised approach to etiotropic treatment of diseases and injuries. Through the use of innovative technologies of optimisation intensity of energy structural capacity, the reliable elimination of energy structural disorders and restoration the biological integrity the bodies of patients will be achieved. This manual is intended for physicians of all specialties.

Introduction

The success of biology together with evolution laws opens up new opportunities in implementing the theory of body integrity in assessment patients clinical state and finding new ways of healing – to treat not a disease but the patient.

The evolutionary essence of living being predetermines the supremacy of energy-structural ideology of medicine, which defines the biological integrity of an organism or body (BIO) as the functional harmony of energy production and energy supply. The main component of BIO is called bio sustainability (BS) – which contemplates living strengths of a body necessary to provide self-resuscitation of cells' DNA (deoxyribonucleic acid) and other organelles, initiate physiological regeneration as well as minimise or eliminate structural deficit.

Therefore, a power of BIO responsible for the energy production multiplied by body cell mass, which we call 'vitalism' is the calculated measure, unique for every single individual. This will set the difference between reliable energy consumption (in active healthy body) and minimum energy used dictated by hypoxic genes just to assure cells survival.

Finally, vitalism is a subject of study of an independent medical science: vitalology. This manual presents the nature of vitalism establishes a new model of healing.

CHAPTER 1

Phenomen of Human Vitalism

1.1 Cellular response to stimulation and alteration

Functioning living cells are the realisation of certain genetic programs in response to numerous external signals. In the same cell these extracellular signals might activate cell mitosis, movement, growth, proliferation or might cause cellular self-destruction or apoptosis. Final cell response (if it is any) depends on concentration of extracellular molecular stimulants, a number of cells receptors and its sensitivity.

By stimulating synthesis and proliferation intracellular organelles a cell or cells undergo adaptation either in a form of hypertrophy (increase in size) or hyperplasia (increase in a number of cells), if talking of a tissue or an organ alone.

Each cell energy production is strictly personalised and supported by genes. It varies with changing cells activity, and suffers from diminished nutrients (proteins, carbohydrates, fat), electrolytes (Na, K, Ca, Cl) and micro-circulation alteration.

Cell/organ function depends on intensity of cellular metabolism which is different at any given time. Therefore, the strength of life ('vitality' or 'vitalism' – a new term) describes a level of intensity of metabolism; or calculated amount of energy immediately produced in active single cell to facilitate effective response to 'fight and flight' mode.

Cell integrity is a minimum level of energy produced that is required for sufficient cell preservation. Plenty of conditions still need to be met for cells to retain its vitalism. If these are not maintained, the living cell is dying.

The conditions are:

1. Energy supply
2. Physicochemical state and membrane structure
3. Enzyme activity
4. Ion balance
5. Genetic factors
6. Energy structural conjugation (vitalism)

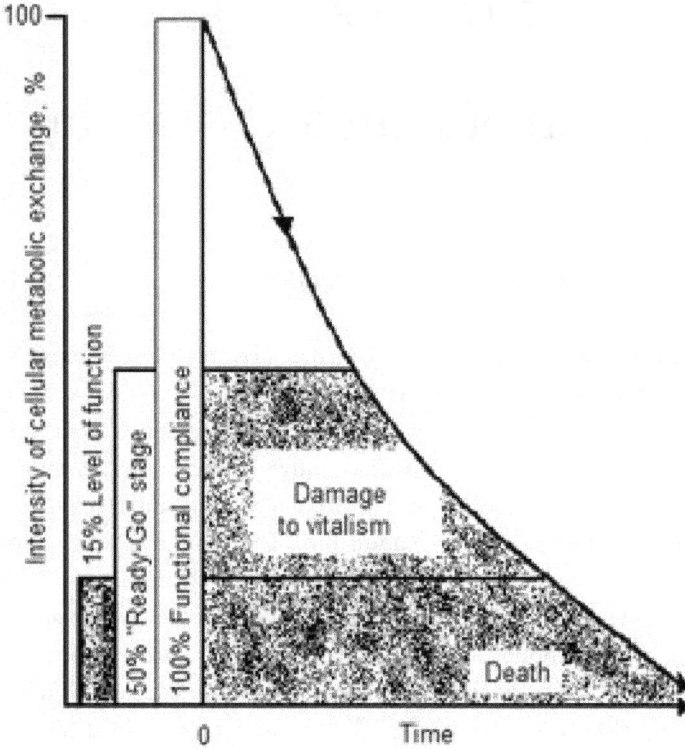

Fig 1.1 The implication of metabolism on an active cell

... Break in energy supply and therefore structural disorders trigger similar changers as an ischemia or re-perfusion syndrome. It affects all stages of energy production cascade:

a) ATF synthesis
b) ATF transportation to target cells
c) ATF utilisation

In tissue hypoxia – the disorders observed mainly on a stage of ATF production – it decreases in a two-phase fashion: initially sharply and then delayed, all due to the depression of glycolysis. Glycolysis alteration contributes to inadequate microcirculation, which in a chain reaction affects further transport and utilisation of ATF

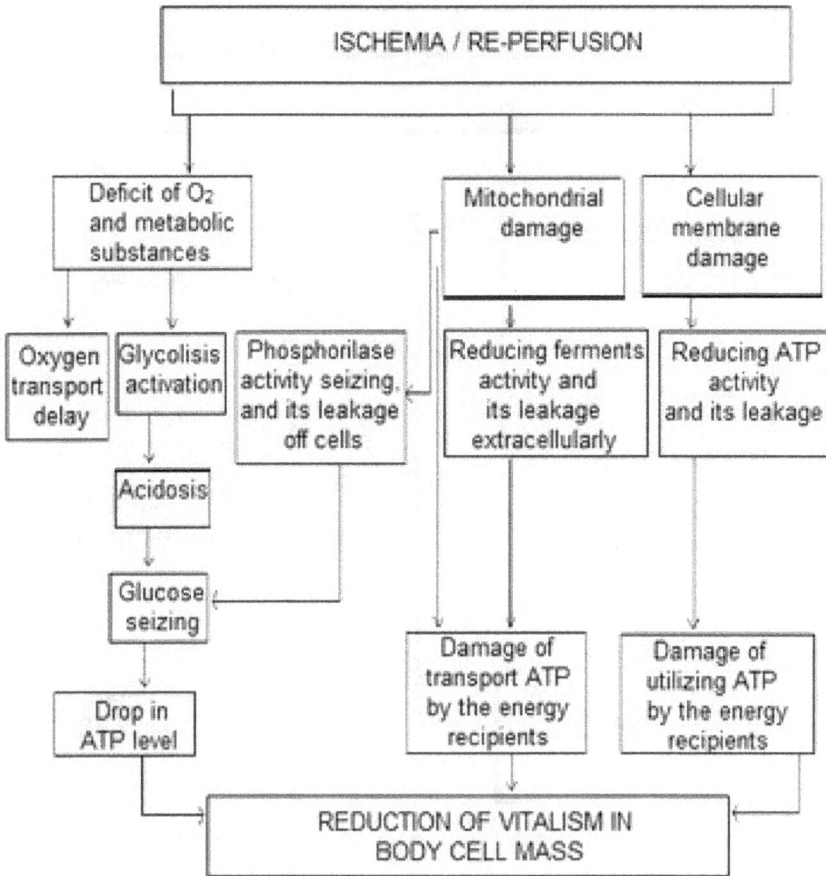

Fig. 1.2 Vitalism disorders in ischemia and re-perfusion

If ischaemia lasts longer than 10 min the adequate energy delivery is preserved. This happens at the expense of further mitochondrial damage (see Pict 1.3)

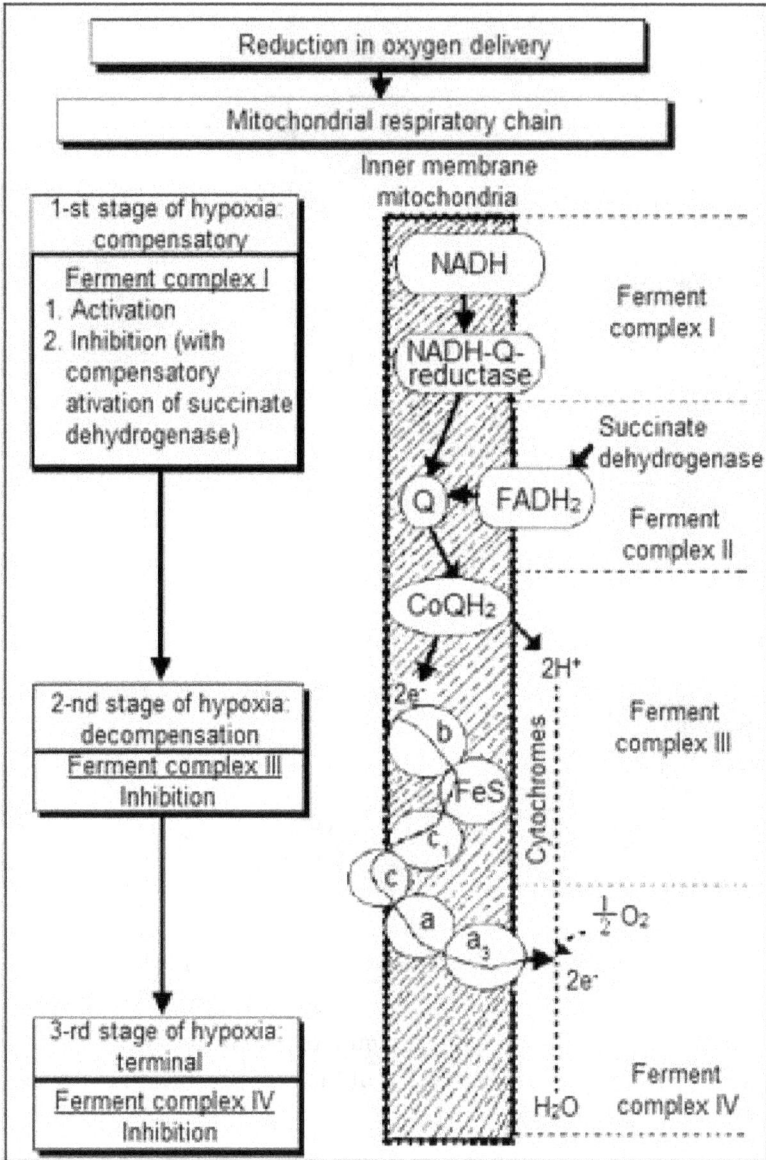

Fig 1.3 Mitochondrial respiratory chain impairment

1.2. Energy- structural conjugation

Conjugation structure and energy determines that a transfer of electrons from carbon and hydrogen to oxygen in the utilisation of nutrients allows these atoms to reach stability. This makes this process energy beneficial. As one atom of oxygen takes two atoms of hydrogen to form a molecule of water, its calculated that each minute in body cells to transport around 2.86×10 electrons and will require in average 264 ml/min of oxygen. The chemical reactions also generate a current of 76A, which is produced and later consumed by an adult at a rate of around 85.88W per minute (G.A. Apanasenko, 1992). It is crucial that this 'biological battery' be able to generate 15-20 times more energy to secure body response to different stress conditions.

The structural basis of vitalism is determined by four groups of cells:

- Cells in apoptosis or necrosis
- Hibernating or stunned cells
- Functional cells of a body, main providers of organs activity
- Regenerated (proliferated cells).

Every minute 1 million cells die and the same number regenerates to maintain energy-structural balance.

The physiological state of a body is the main component of the body's life source, necessary to cover basic body's needs: growth, proliferation, regeneration, multiplication and finally death.

There is no central energy storage that can meet all cell requirements in the body. Therefore, each cell must allocate where the limited energy resources will be directed. As oxygen demand increases, cells can produce more energy; by speeding up metabolism and initiate regeneration process when the energy resources are depleted. While at rest this self-regulating process goes into hibernation.

Adenyl nucleotides such as ATP, ADP and cyclic AMP that play the leading role in metabolic bio cycles are also necessary for cells reparation and regeneration. A number of other organic compounds like diphosphate (pyrophosphate) play a part as an energy buffer in muscle cells, as an additional phosphate group to different biological compounds, and the raise sensitivity of numbers of metabolic cascades.

1.3 Major Energy Recipients

There are five main energy distribution pathways. All of these are design to various chemical body reactions, maintain osmotic cells and tissue integrity as well as contributing to a body's movements or heat production.

ATP is a universal converter of preserved energy into one of processes as it shown below:

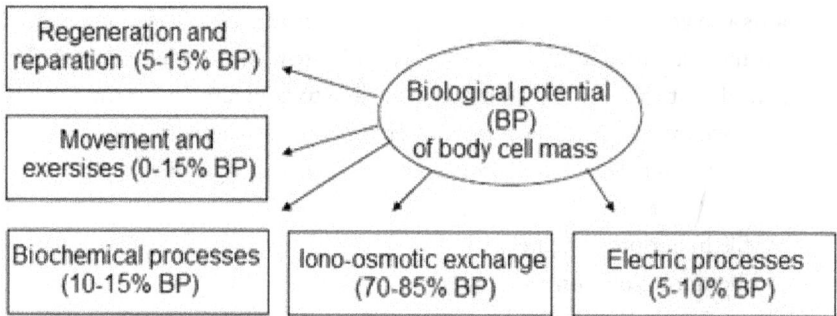

Graph 1.4. Mass body cell (MBC) potency

In addition, other tree phosphates such as guanosine triphosphate, uridine triphosphate and cytosine triphosphate are main ingredients for the synthesis of proteins, carbohydrate and phospholipids. Depending on availability, intracellular ATP, ADP, AMP, mitochondria, etc. can initiate different metabolic pathways.

For instance, in active mitochondria there will be enough substances for phosphorylation in a form of ATP, PK, NTP, and oxygen will be actively absorbed as phosphorylation cannot occur without oxidation. While in resting cell mitochondria in saving mode with oxygen and plenty components for an oxidation, but lack of ADP makes phosphorylation process impossible. In later cases, cell respiration rate decreased while the concentration of energy transporters like NADH increased. It is calculated that in cells a speed oxidative phosphorylation depends on the ATP/ADP ratio.

Increasing energy demand leads to mitochondria 'awakening' and thermodynamic regulation of cells respiration becomes kinetic. Under this condition, the equilibrium of a self-reparation dynamic system (SRDS) depends on cell's energy charge (Qv) and can be expressed by the equation:

$$Qv= (ATP + 0.5ADP)/(ATP + ADP + AMP)$$

The leading part of 'energy charge' is maintained by intracellular metabolic reactions. SRDS occurrence is highly specific: for a sample for a cell reparation, it is only its own free energy (ATP) used, no 'infused into vessels' (exogenous) ATP can help, as it is not recognised by mitochondria.

This makes SRDS significant and multifaceted. In some conditions it catalyses many biological reactions, while different scenarios cause its inhibition. The role of ATP, ADP and AMP cannot be overlooked: these are main components of cell respiration. A no less important component is oxygen, as in hypoxia SRDS is grossly compromised.

A fall in vitalism due to any damage in tissue integrity leads to building up energy-structural dysbiosis (ESD) which interfere with dynamic self-reparation.

The consequence of this is a decrease in biological resistance and an increase cells vulnerability.

In hypoxia, where oxygen and energy deficiency present, the biological integrity of an organism (BIO) can be represented by an equation:

BIO = CBM (cells body mass) x vitalism

This relationship explains that vitalism fully ensures the biological body's integrity and maintains body life forces. When lacking oxygen and energy components in gradient hypoxia there is a tension which occurs in the whole system – acidosis persists, and if conditions do not improve it will trigger cellular necrosis (apoptosis). Acidotic alteration (AA) is a universally known phenomenon of cell death.

In order to enhance vitalism, you have to act on all three different levels:

- molecular (amplification, i.e., increasing the number of identical genes; polytenization of chromosomes (forming cable-like structures); and synthesis of adaptation enzymes
- cell organelles (hyperplasia, multiplication)
- tissue and organs (proportional increase in functional capacity, hypertrophy)

1.4 Implementation Of Vitalism

Maintaining the body's compatibility is one of the conditions in general adaptation syndrome (GAS) controlled by central nervous system (CNS) through polyglandular somatotrophic response.

An insufficient volume of the diffusion surface areas of transcapillary exchange (SATE) diminishes resource supply for BCM (body cell mass), compromising its transportation through microcirculation, which is normally done by one of three processes: diffusion (active and passive), osmosis and active transport.

The oxygen supply is determined by:

1. Oxygen content in arterial blood
2. Blood flow velocity

High intensity oxygen transport in tissue can cover for insufficient SATE and associated bioenergy deficiency (BED).

Global oxygen delivery (DO2) can be described by an equation:

$DO2 = a\text{-}vO2/Ht$ (ml/l), where a-vO2 is arteriovenous difference in the oxygen content and Ht – haematocrit (l/L)

A-vO2 is a function of oxygen capacity, while Ht is set by cardiac output application on a cross-session of vasculature, and highly dependable on tissue perfusion.

The estimated figure of DO2 for men is around 116 ml/l and for females 128ml/l

Endothelium (EC) is specific organ which distribute the oxygen and nutriments according to the needs of BCM. The cells of EC are responsible for the secretion of active substances into the vascular muscle wall or directly to a bloodstream.

Vasodilatation is caused by:

- nitric oxide (NO)
- prostacyclin
- vascular endothelial growth factor (VAGF)

They form complex linked system controlling the structure, tone and haemostatic properties of the vascular wall, initiating the appropriate response in the vessels.

Factors that contribute to vasoconstriction:

- arachidonic acid and its derivates (prostaglandin H2 and thromboxane A2)
- free oxygen radicals and elevated level of endogenic peroxides

- a high level of angiotensin-1 (AT1)
- peptides of endothelium , of which most studied endothelin-1, which causes sustainable constriction and smooth muscles proliferation

Disbalance in ratio endothelial vasodilators and vasoconstrictors lead to endothelial dysfunction.

The main prognostic factors of microcirculatory-mitochondrial distress syndrome (MMDS) are:

- reduction of NO, a drop in prostacyclin
- increase in free radicals, peroxides, AT1 and VAGF synthesis
- reduced sensitivity of smooth-muscular cells to vasodilators

In the end, the fate of somatic cells predetermined by a state of endothelium, which in severe energy and oxygen deficiency induce cell apoptosis or necrosis.

Therefore, the endothelial damage inevitably interferes with the nosological vitalism of a body.

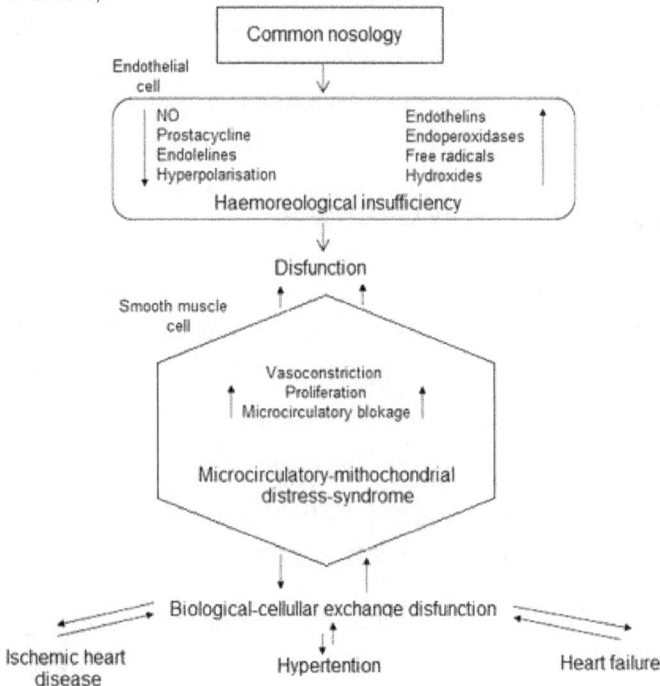

Main Endothelial Factors in MMDS

The organ/cell function at any moment of time corresponds to a strictly equivalent number of structures 'producing' equal amounts of biological potential (BP).

This can be achieved in different ways. First is by constantly fluctuating the number of actively functioning structures. With higher demands in functional load, the number structures will increase, and in the opposite case, with metabolic processes slowing down only some parts of the organelles will activate.

This principle named: 'The principle of asynchronous operations '. It works in all levels of an organism's hierarchy: from cellular to systemic and from systemic to organismic level.

In continuous exercise routine, the number of active structures of a body generating energy becomes exhausted. Two other processes come into play: increased cells proliferation (a rise in number) or cells hyperplasia (an increase in size). Alongside this, is the speeding up of reparative regeneration (restoration in number, replacing apoptotic cells).

The ideology of constant structural balance is not limited to a work of different group of cells only. It is also true that the same kind of cell performs not one but a few functions.

This undoubtedly expands 'material base' of vitalism.

1.5 Biological and Energy Quantums.

Biological Cycle Quantum

Maintaining the integrity of the biological cycle quantum (BCM) requires a strictly complementary energy supply for physiological processes and reparative regeneration.

This agrees with the systemic quantisation of life support. The meaning of this is that the components of biological and energy-dynamic activities or values can be converted into a range of finite discreet values.

Therefore, the minimum biological potential could be defined as the biological quantum (BQ).

It is fascinating that the specific biological quantum (SBQ) of an older person is only 13% less than the same in a young one. However, there is greater gender difference, in that females are superior to males with a margin of 33% whether old or young individuals.

In this sense, SBP is holistic and undividable. It's the quality which can secure the body integrity in all stages of ontogenesis. According to Rubner, in the human body only 5% of BQ is spent on regeneration, while animals will deviate for 35% of SBQ. The high energy efficiency of reparative processes in humans can be accounted for by the eternal constancy BP throughout a lifetime.

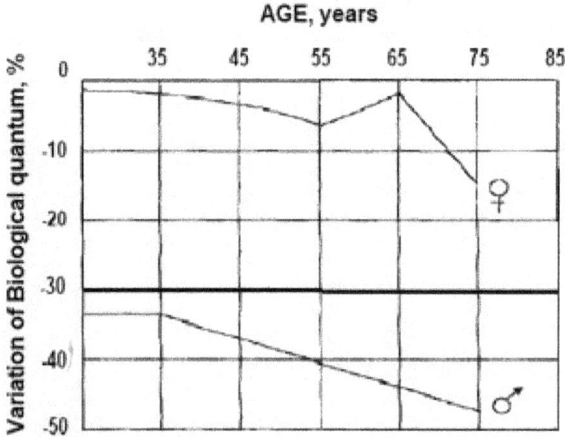

Fig 1.5 Age-related changers in the specific biological quantums

Hence, SBQ has such a nature of integrity that it cannot be extracted from separate elements. Furthermore, it cannot be considered as mathematical sum of organic components. Its unique in all its manifestations in males and females.

1.5 2. Energy Protective Quantum

According to Rubner, all types of mammal species inherit the same size energy fund, which is an average 725800kcal/kg (generated over a lifetime) – four times more than any other living species can do. Why is the human power such a striking exception? Only because SBQ is secured by more advanced pathways of oxygen and energy delivery and consumption.

The type of oxygen and energy transport that does not limit the achievement of the maximum value of SBQ for BCM we name the energy protective quantum (EPQ) By enhancing its own EPO with the additional storage of energy and substances the body increase its assets and secures a degree of own reliability.

This distinguishes live forms from non-living subjects. For the latter, life support is an act of dissipation (a process of slow vanishing), the gradual involution to a state of equilibrium. Mammals, on the contrary, express a resistance to dissipation, ensuring stable evolution even in extreme disequilibrium, providing EPQ guarantees the full power of BQ.

If we imagine CBM as a sphere, then its volume (V), where all biological cycles are taking places can be estimated by the equation:

$$V = \tfrac{4}{3} \times \pi\ R^2$$
$$R - \text{sphere radius}$$

The surface area (S) then can be equal:

$$S = 4 \times \pi R^2$$

And the ratio volume to surface area (V/S):

$$V/S = R/3$$

Its poor geometrical progression: if a radius of a sphere increases twice, the surface area grows by four times and the volume of a sphere increases by eight times. As a result, the interchange between CBM and an environment in each quantum biological cycle (QBC) will become less and less effective. And it might come to a case when constantly increasing mass will not be able to obtain the necessary quantity of oxygen and substances.

At the same time, the removal of by-products (acidic components) from cells may be significantly diminished for each QBC. It would have undoubtedly caused cell apoptosis, if nature had not found a way out and resolved it by simple cells division and multiplication.

In addition, it should be stressed that a trigger is not reaching a critical point of increasing body cell mass (BCM) but achieving a dangerous ratio between BCM volume and its surface area, covering a mass. If proliferation is suppressed, the morpho-structural organisation is disturbed. The balance becomes negative due to an energy deficiency that induces cell apoptosis or necrosis.

Established thermodynamic equilibrium in QBC corresponds with Le Chatelier's Principle: if a dynamic equilibrium is disturbed by changing the conditions, the position of equilibrium moves to counteract the change.

As the second thermodynamic law ensures entropy (a measure of clutter), the QBC ensures the body's constant battle with its increase, providing energy-protective conjugations, aka vitalism.

The relationship between BQ and EPQ in reaching QBC is presented in Fig. 1.7:

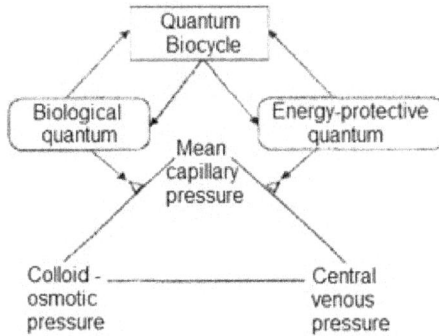

Fig 1.6 Quantum bio-cycle

The aetiology of different diseases can disturb the compliance BQ and ERQ due to hypovolemic hypoxia. However, with decreasing blood flow, BP and a dropping in intensity of the biological cycle, the compensation can be reached by increasing the time of erythrocyte deoxygenation in capillaries. This adaptation cycle, limited by minimal arterial pressure, requires for plasma between vessels and tissues. As opposed to this, in the case of arterial hypertension the rate of erythrocyte deoxygenation decreases, regulating the biological quantum cycle.

The milestone of circulation is venous return with its velocity directly proportional to the gradient Pc/CVP (capillary pressure/central venous pressure), which defines the relationship between blood volume and the capacity of the vascular channel. Change in cardiac index (CI) with fluctuating venous return is possible due to the dynamic increasing and decreasing relaxation of the myocardium. The regulation of CI grossly depends on right atrial pressure, which is known as CVP. All these mechanisms are responsible for the maintenance of EPQ in nosologically changing BC.

A distinctive and dangerous feature of haemodynamics in disease is an increase in the resistance of pulmonary circulation. A hyperdynamic state of circulation triggered by sympathetic adrenal response is a crucial in order to achieve increasing EPQ.

Stimulation of the cardiovascular system will increase oxygen consumption, but at the same time the amount of oxygen available for each litre of blood will drop due to the incomplete deoxygenation of erythrocytes. The result: exacerbated tissue hypoxia. To summarise: the excessive increase of CI is dangerous as it can compromise vitalism.

1.6 Energy-Protective Theory

A possible dangerous outcome of vitalism is insufficiency caused by a disbalance of energy-structural coupling mechanisms. Thus, if not enough energy is produced, the concentration of intracellular H+ (hydrogen) ions grows, leading to osmotic leaking of extracellular fluid. This triggers damage to the compartments, and a progressive increase in intracellular osmolality (it could reach 40 mosm/l in just 10 min). If such changes take place in the myocardium, it could cause a drop in CI, initiating cardiogenic shock. These ion-osmotic disturbances can lead to organ or multi-organ failures.

That is why the role of peri-nosological monitoring is the prevention of minimal discrepancies in osmolality intra and extracellular compartments, which will limit or regulate the energy requirement.

The Energy – Osmotic Relationship is best described by the equation:

$$OP = 332-0.026X\ DO2 - 0.137 \times VO2$$

where:

OP – Plasma osmolality
DO2 – Oxygen delivery (ml/min x meter square)
VO2 – Oxygen consumption (ml/min x meter square)

As a result ,if oxygen consumption decreases the inner body environment becomes hyperosmic.

This makes osmolality one of the important of vitalism's parameters.

The fundamental reason for the vulnerability of any living system is the discretion of living in the modern world, meaning there is a limitation in inner evolution of a species in order to maintain distinctive living system characteristics. Self-renovation and reparation are not enough to resist all external interventions. For a sample to withstand the second thermodynamic law is only possible by employing external influences, which given time will lead to a system evolution, changing its structure, rather than reaching its stabilisation. The evidence of later could be an existence of several types of biological integrity of organism. These can be manifested by numerous conditions of disturbances in vital resistance of a body.

The principles of energy biometrics allow the estimation of all biological processes, for instance biological potential (BP), changing with the intensity of a metabolism or with consumption of oxygen or energy substrates. Therefore, the current BP can be calculated by comparing the current metabolism (CM) level with the impersonated one, which involves the readiness for activation all compensatory body mechanisms, called proper basic metabolism (PBM). This could serve as a measure of a standard of a body BP.

The quantitative value of BP is determined by the ratio of CM to the PBM as a percentage:

$$BP = 100 \times CM/PBM$$

BP (%)	Level BP	Energy consumption (kkal/kDj×24h)	EPS (%)
85-147	Hypobiothy	661/2763/BS - 1165/4868/BS	Hypo-energybiothy
148-192	Normobiothy	1166/4869/BS - 1375/5746/BS	Normobiothy
≥ 193	Hyperbiothy	≥ 1376/5747/BS	Hyper-energybiothy
≤ 84	Pathobiothy	≤ 660/2762/BS	Patho-energybiothy

where BS – Body Surface

Table 1.1 Biometric assessment of BP and Energy-Structural Status (ESS)

Where BS – Body Surface, metres square
BP, % BP level EPS/CM (kcal/kg x day) EPS evaluation
85–147 Hypobiotic 661/763 x BS – Hypoenergy state
1165/4868 x BS
148–192 Normobiotic 1166/4869 x BS Normoenergy state
1376/5746 x BS
193 and > Hyperbiotic 1378/5747 x BS and > Hyperenergy state
84 and < Pathobiotic 660/2762 x BS Pathoenergy state

The principles of formation of a body's BP are universal, and work for all types of levels of its coordination. Systemic delivery oxygen and energy substrates ensure the body's vitalism, providing close link between energy-protective status (EPS) and oxygen delivery (DO2)

DO2 can be calculated by:

$$DO2 = CaO2 \times CI, \text{ ml/min/m}^2$$

CaO2 – oxygen content of arterial blood (ml/l)
CI – Cardiac Index 1/min x m^2
DO2 fluctuates between 493 and 582 ml/min x m2 if the integrity of BCM is maintained and EPS is in the Normoenergy state. Current energy protective potential (EPP) can be calculated using proper basic EPP.

The latter is the amount of oxygen substrates that guarantee constant energy delivery into BCM, necessary for EPS to adapt to any conditions.

Since in order to increase the DO2 more than 3.2 times, BCM should develop oxygen insufficiency (to facilitate a demand), therefore the estimated DO2 will equal:

$$DO2 = 3.2 \times \text{Current VO2 (ml/min/x m (square))}$$

And the fluctuations of this are about: EPS = 100 x DO2/estimated DO2 (in %)

The algorithm of the relationship between EPP, DO2 and EPP (in %) can be seen in this table:

EPP (%)	VO$_2$ (ml/min/square meter)	Level of EPP
85-147	331-492	Hypo-energyprotective
148-192	493-582	Normo-energyprotective
≥ 193	≥ 583	Hyper-energyprotective
≤ 84	≤ 330	Patho-energyprotective

Table: 1.2 Biometrics of Energy- Protective Potential (EPP)

One of the properties of biological integrity of a body is the systemic sequential quantisation of vital activity. It means that all processes constituting biological potential (BP) and energy protective potential (EPP) can be divided in their continuum on to discrete integrated quants, which cooperate.

The minimised BP is the biological quantum (BQ) which can be estimated by equation:

$$BQ = OP/BM$$

where:
OP – oxygen pulse, mlO2/heart contractions
BM – Body mass, kg

Knowing the ratio between OP and BCM (body cell mass) you can calculate the basic biological quantum, mlO2/kg.

In order to estimate BCM there are few parameters need to be considered: body weight (BW), height (H) and age (A).

Also, the proper value of body mass cell (PBMC) can be calculated.

Thus: Men under 35 years old:

$$BCM = 1.7954 + 0.4497 \times BW + 0.7014 \times H$$

$$PBMC = 43 + -0.85, kg$$

Men 35–45 y.o: $BCM = -2.1260 + 0,4459 \times BW + 0.7585 \times H$

$$PMBC = 41.6 + or - 0.62, kg$$

Men 45 and > $BCM = -0.2091 - 0.4420 \times A + 0.4449 \times BW + 0.6426 \times H$

$$PBCM = 35.6 +/- 1.42, kg$$

Women under 35 y.o: $BCM = -3.0545 + 0.3901 \times BW + 0.9385 \times h$

$$PBCM = 27.0 +/- 1.35, kg$$

Women 35–45 y.o: $BCM = -2.6592 + 0.3465 \times BW + 0.8911 \times H$

$$PBCM = 26.5 +/- 0.79, kg$$

Women over 45 y.o: $BCM = -0.9042 - 0.3048 \times A + 0.39 \times BW + 0.7417 \times H$

$$PBCM = 25.0 +/- 0.37, kg$$

By comparing the values of BCM and PBCM we can judge more precisely the person's energy-structural constitution, consistent with individual BQ and specific biological quantum (SBQ).

How the latter fluctuates according to an age and gender you can see in the Table: 1.3

Gender/Age	Biological Quantum × 10² (mlO₂/kg)	Specific Biological Quantum × 10² (mlO₂/kg)
Females up to 35 y.o.	5.97 - 4.36 (5.17)	12.62 - 9.20 (10.91)
36-45	5.89 - 4.31 (5.10)	12.42 - 9.09 (10.76)
46-55	4.60 - 3.62 (4.11)	11.59 - 9.12 (10.35)
56-65	4.21 - 3.60 (3.91)	11.28 - 9.65 (10.48)
66-75	4.1 - 2.88 (3.49)	11.15 - 7.83 (9.49)
All ages	4.95 - 3.75 (4.35)	11.81 - 8.98 (10.04)
Males up to 35 y.o.	4.82 - 3.5 (4.16)	8.48 - 6.16 (7.32)
36-45	4.6 - 3.18 (3.89)	8.46 - 5.85 (7.16)
46-55	4.2 - 3.14 (3.67)	7.73 - 5.78 (6.75)
56-65	4.2 - 3.00 (3.6)	7.70 - 5.55 (6.63)
66-75	4.07 - 2.70 (3.38)	7.68 - 5.10 (6.39)
All ages	4.38 - 3.1 (3.74)	8.01 - 5.69 (6.85)
Average	4.46 - 3.42 (4.04)	9.79 - 7.33 (8.56)

Table 1.3: Estimated values of biological quantum (BQ) and specific biological quantum (SBQ)

Biological quantisation gives us an opportunity to appreciate the phenomenal personification of each biological system. Individual assessment becomes clearer if compare the deviation of present values (in real life) from the estimated ones. An analysis of the latter indicates BQ is less informative than SBQ. Although the absolute and relative changes in BQ of males and females as they aged are more prominent than values of SBQ, only SBQ can reflect the true BP of each unit of BCM. Therefore, the mathematical evaluation and confidence intervals of SBQ can characterise how a unit of BCM is capable to perform all sorts of functional activities to facilitate BP.

Comparison of the individual values of SQP with proper values of such estimates with great accuracy the true biological age of a person.

A rapid decline in BQP testifies the critical life's activities disorder, caused by a vitalism deficiency. The discreteness of the oxygen and energy-subtracts transportation, which does not compromise the full achievement of SBQ for BCM in the individual life quantum we classify as energy-protective.

The structure that ensures the complementarity of the energy-protective quantum is shown in

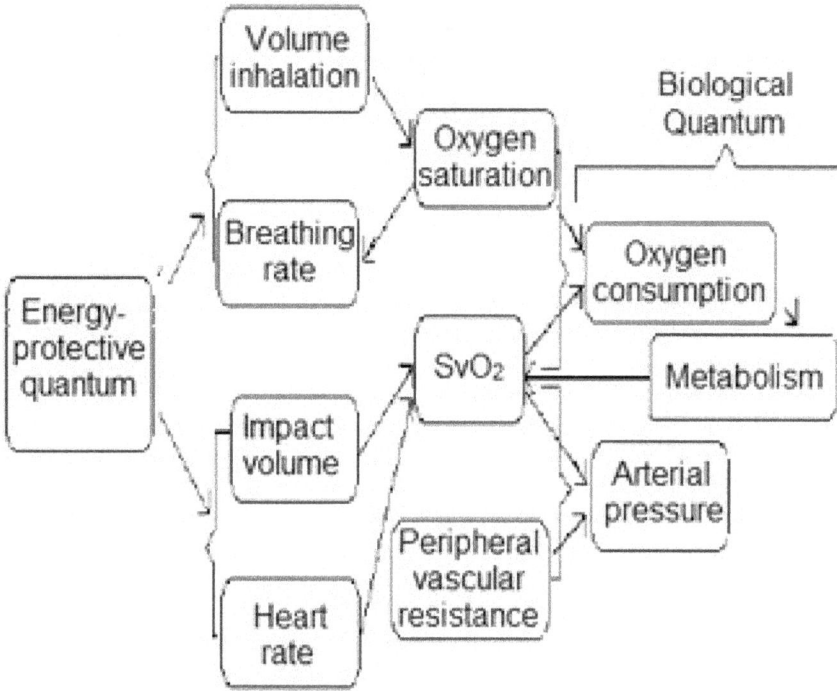

Fig. 1.4 Complementarity of Energy-protective Quantum (EPQ)

Due to the rhythmic correspondence of singular EPQ to each BQ, the degree of ordered structural organisation crucial to an integrity preservation is reached.

Therefore, the specific energy-protective quantum (SEPQ) is determined by the equation:

$$\textbf{SEPQ = SBQ/ (CaO2 – CvO2)/CaO2}$$

where:

(CaO2–CvO2)/CaO2 – Oxygen content of a blood

The CONFIDENCE INTERVALS and MATHEMATICAL EVALUATION of EPQ and SEPQ, mlO2/kg are as follows:

23

Gender/Age	Energy-protective Quantum × 10^2 (mlO₂/kg)	Specific Energy-protective Quantum × 10^2 (mlO₂/kg)
Females up to 35 y.o.	20,58 - 15,03 (17,80)	43,52 - 31,72 (37,62)
36-45	20,31 - 14,86 (17,58)	42,83 - 31,34 (37,01)
46-55	15,86 - 12,48 (14,17)	39,96 - 31,44 (35,70)
56-65	14,52 - 12,41 (13,46)	38,89 - 33,28 (36,08)
66-75	12,42 - 10,67 (11,55)	33,75 - 28,99 (31,37)
All ages	16,74 - 13,09 (14,92)	39,79 - 31,35 (35,50)
Males up to 35 y.o.	17,21 - 12,50 (14,86)	30,29 - 22,0 (26,14)
36-45	16,43 - 11,37 (13,90)	30,21 - 20,89 (25,55)
46-55	14,48 - 11,21 (12,85)	26,65 - 20,64 (23,65)
56-65	14,48 - 10,34 (12,41)	26,81 - 18,94 (22,88)
66-75	13,57 - 9,0 (11,29)	25,6 - 17,0 (21,30)
All ages	15,23 - 10,88 (13,06)	27,91 - 19,89(23,90)
Average	15,99 - 11,99 (13,99)	33,85 - 25,62 (29,74)

Fig 1.5 Estimated values of EPP and Specific EPP (SEPP)

Quantum (SEPQ)

The relationship of BP in the quantum biocycle depends on coefficient of conjugation (CC). This allows us to judge the nature of thermodynamic changers in each quantum biocycle.

Here shown how CC is varies with a gender and an age:

Gender/Age	Coefficient Conjugation / Specific Coefficient Conjugation, c.u.
Females up to 35 y.o.	4,73 - 2,51 (3,62)
36-45	4,71 - 2,52 (3,61)
46-55	4,38 - 2,71 (3,55)
56-65	4,03 - 2,95 (3,49)
66-75	4,31 - 2,60 (3,45)
All ages	4,43 - 2,66 (3,54)
Males up to 35 y.o.	4,92 - 2,59 (3,76)
36-45	5,16 - 2,47 (3,82)
46-55	4,61 - 2,67 (3,64)
56-65	4,83 - 2,46 (3,65)
66-75	5,02 - 2,21 (3,62)
All ages	4,92 - 2,48 (3,70)
Average	4,67 - 2,57 (3,62)

Fig 1.6 Estimated values of Coefficient Conjugation

The real coupling conjugation factor (RCC) can be determined by a formula:

RCC = REPQ/SBQ

Where: REPQ – real energy-protective quantum

In the case of energy deficiency, the maintenance entropy depends on the relationship between real oxygen status (ROS) and estimated oxygen status (EOS).

As opposed to this, with no lack of energy:

ROS/EOS = 1

This declares a state of thermodynamic equilibrium established in QBC.

The duration of the quantum biological cycle (QBC) fluctuates rhythmically within every 50+/-4 repetitions. It corresponds with basic metabolic processes.

For instance, the protein secretion takes on average 40–60 minutes. The periodicity of changes falls into "repetitive mode" movements.

The integrity of BCM is kept unaltered due to continuous transformations of structure-functional and time-isometric discreteness.

Therefore, the fluctuations of SEPQ and SBQ should remain within the normal cycle. Otherwise, the rapid exhaustion of the reserve and lack of a synthesis of BCM units will re-form QBC aimed at preservation of QBC and neutralisation of damaging effects.

Gender/Age	Specific Quantum Bicycle, sec.
Females up to 35 y.o.	0.78 - 0.58 (0.68)
36-45	0.75 - 0.55 (0.65)
46-55	0.67 - 0.57 (0.62)
56-65	0.63 - 0.54 (0.59)
66-75	0.63 - 0.54 (0.59)
All ages	0.69 - 0.55 (0.62)
Males up to 35 y.o.	0.84 - 0.59 (0.72)
36-45	0.79 - 0.58 (0.68)
46-55	0.71 - 0.54 (0.62)
56-65	0.71 - 0.53 (0.62)
66-75	0.71 - 0.50 (0.61)
All ages	0.75 - 0.55 (0.65)
Average	0.72 - 0.55 (0.63)

Table 1.7 QBS regarding the age and a gender of an individual

Despite the relatively stable SQB, the QBC varies following fluctuations of the SEPQ. Its interaction creates a symmetry in QBC and the structural functional body's response as a whole.

The robustness of the complementariness of SQB and SEPQ can be estimated by calculating a correlation coefficient during 6–10 QBC. This coefficient is used quite often to characterise the interaction force.

It works in linear and non-linear cases, relying on the linear approximation of a process under study.

The quantum biocycle might fluctuate as it shown in a Fig. 1.7

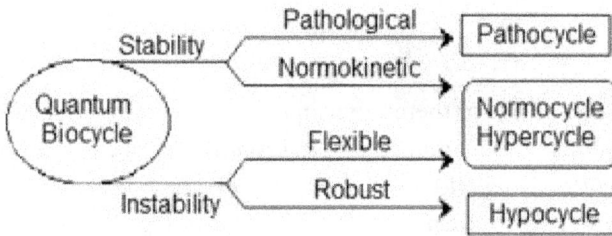

Fig 1.8 The liability of quantum biocycle (QB)

As shown in the graph, if the relation between SQC and SEPQ is weak the QBC holds homeokinetic properties. In a situation when correlation is strong, it can turn the homeokinetic stability of QBC into a pathological one, reversing normocycle into pathocycle.

The value of the linear regression coefficient (R) allows us to estimate QBC variability at all fluctuations SQC and SEPQ. If 'R' is heading to a '0' value, then the QBC variability is flexible, if 'R' closer to '1', it becomes difficult, and it threatens to convert normocycle into hypocycle. Monitoring body vitalism and reaching QBC recovery requires the maintenance of the energy supply, primarily the process of auto-renewal in BCM. For this purpose, it is necessary to stick to the normo energo-biotic intensity of metabolism. This will be possible with precisely provided amounts of oxygen and energy resources, to avoid energy starvation.

To distinguish the character and severity of structural damage which occur at an imbalance in inflammation and regeneration systems, we are presenting the lymphocytic monocytic morphogenic Index (LMMI), which is equal to:

$$LMMI = (L/(L + M)) \times 100$$

where:
L – lymphocytes number, 10x9/L
M – monocytes number in a blood film

Dropping the LMMI to 75% and below, thus manifesting lymphocytic-morphogenic deficiency, aggravates nozogenous disorders.

The regulatory role of the immune system is to balance pro-inflammatory response on different tissue cells and the ability to regenerate. When the reduction in LMMI reaches 50%, the instability and destructiveness trigger irreversible processes. In such reparation distress, the structural deficit became almost insurmountable.

Systemic intravascular alteration can be evaluated by the intra vascular alteration (IVA), as shown below:

$$\text{IVA} = 100 \times ((M + ND) + L)$$

where:

M – Monocytes number

ND- non-differentiated lymphocytes

L – Leukocytes number

If the IVA is over 110% then alteration is severe and it is the pro-coagulative circulation type, resulting in plasma 'leakage' and microcirculation blockade.

The treatment will require all energy-protective technologies of perinozological support.

From the point of view of dynamic vitalometry, it is useful to monitor the level of all functional reserves. In order to predict and pre-empt such vascular damage it is advisable to monitor the ratio of systolic/diastolic pressures and pulse pressure variations.

Calculation of the harmonic systolic, diastolic and pulse blood pressures are measured in comparison with the absolute harmony number (0.618) which allows us to estimate the danger of an exhausted energy supply in the range of real values of blood pressures.

The violation of the peripheral nervous system can be determined by the peripheral index (PI), which is the reflection of interaction of and organ's perfusion pressure (diastolic DBP) and HR.

$$\text{PI} = (1 - (DBP/HR)) \times 100$$

With a PI from 1–7 we talk about euthonia, while a PI on the rise with an activation sympatho-adrenal mechanism, and its negative in vagotonia.

Therefore, monitoring of vital stability reserves and an ability to tackle a body's adaptability gives us a chance to achieve the security of peri-nosological support.

CHAPTER 2

Logic of Vitalism Concept

2.1 Auto regulation of adaptation

Living organisms adapt to a constantly changing environment. Without this, it would be impossible to survive, as well as to function in condition like extreme temperatures, pressures, altitudes, hypoxia, weightlessness and infections.

Adaptation is of great importance to humans and animals. It allows us not just to maintain an existence in a challenging environment but make active adjustments of our body's physiological functions and behaviour in accordance with changes in conditions, sometimes even before they change.

Because of adaptation, the consistency of a body's inner self, such a blood, its components, acid-alkaline state, osmolality and so on is maintained. In case of prolonged exposure to damaging environmental factors, significant deviation from the norm may occur, sometimes beyond a safe level, which leads to functional organ disorders, and the manifesting of pathological symptoms.

The role of adaptation not only maintains a consistent body inner state, but also rearranges different functions of an organism to ensure it overcomes unexpected physical, emotional and other stress.

In addition, an adaptation can lead to behavioural changers, which is particular in evidence in the animal world (sample: hibernation). In an adaptation of highly advance species not only the CNS participates, but a big part played by sympatho-adrenal and hypothalamo-hypophisal systems.

When unfavourable conditions arise, an adaptation contributes to the triggering of various bodily mechanisms –protective ones that are unable to counteract the impairment included.

The state of physiological adaptation is judged by a sensitivity threshold analysed in a system, which is highly dependent on changes in stimulus intensity.

Cellular adaptation is the cells adjustment to the changing environment, aimed at survival and reproduction.

In advanced species of animals, or plants, an adaptation tends to take place at whole level of an organism, one component of which will be cell adaptation.

Cell adaptations conditionally divided into genotypical and phenotypical. The first, genotypical, occurs due to the careful selection of particular genotype, securing its endurance. On the other hand, phenotypical manifests as a protective reaction from damaging influence. The timing and the intensity of damaging conditions play major roles. The strong hold of the intruder's influence may kill cells or an organism before it adapts. In weak stimulus, or in slow rise of it, so-called cross-resistance develops – cells will become less sensitive not only to damaging factor but to other stimulants too.

The stability caused by a mild irritant can be maintained even after its action ceases. It confirmed on repeatedly used the same agent. However, the rate of cell resistance as well as the duration of an adaptation various greatly.

The degree of cell adaptation, whether in an increase or decrease of the sensitivity threshold, ensures the safe level of functioning (receptor's functioning).

The mechanisms, underlying adaptation depend on the nature of the cells and the type of damaging factor.

The science looking into adaptive response of cells or a multicellular organism (dependably on its organisation) is called biophysics, and it sees an organism as a system opened to an external environment or freely exchanging with its energy and substances.

In this context, the dynamic equilibrium of the process – an inflow and outflow of matter and energy – provides the necessary level of the stationary state of a living system, the permanence of internal constitution and various gradients on its borders. This determines the normal cells or organism functioning under changing conditions. And for controlling an adaptation processes the principles of feedback are there, similar to the endocrine glands one.

2.2 Stress – Response to Extreme Conditions

Under extreme conditions, where adaptation and compensatory mechanisms are not fully effective, shifts in basic homeostatic parameters are prolonged and even progressed. These impairments, in a way, increase and prolong the effect of primary exposure, playing the role of a powerful stress stimulus.

The general rules of stress development on extreme factors influence are linked to the stimulation of the hypothalamo-hypophyseal and sympatho-adrenal systems. The first mechanism causes CNS activation and the triggering of the adrenocorticotropic function of a hypophysis. It has been established that in extreme conditions, first and foremost, the activity of a cortex and parts of limbic system responsible for negative feedback is depressed. As a result, there is persistent excitation of hypothalamic structures, and an increased production of Corticoliberin (CL).

Another effect of high activity of the hypothalamo-hypophyseal-adrenal (HHA) system is the accumulation in blood and tissues stress by-products: catecholamines, serotonin, histamine, vasopressin, kinins, prostaglandins and other biologically active substances, which can stimulate the CNS and PNS links of the hypothalamo-pituitary-adrenal system. In this case, the negative feedback is not strong enough, as a result of which hypothalamus expresses low sensitivity to the corticosteroids depressive influence. This leads to the escalation of the corticosteroids blood level which switches the negative feedback mechanism to 'delayed type'action. The later has long latency period and during it the feedback in the HHA system is absent. Moreover, in these conditions, the positive feedback may be formed through the hippocampus, or some precursors of corticosteroids or its by-products, accumulated in stress to keep the hypothalamo-pituitary-adrenal system stimulated, contributing to a chain reaction.

Thus, the intensity, duration and severity of stress which manifests in extreme conditions is determined by the prevalence of stimulating and potentiating effects on the HHA system, which eventually gets overstimulated. The hyperactivity is expressed in a synchronised and extremely sharp increase in the functioning of the central, peripheral and transport reserve links of a system, responsible for synthesis, secretion and transport hormones. This system also contributes to a sensitivity increase in stimulating factors if it is suppressed or inhibited. Therefore, the stress transformation of the HHA system is rather dangerous, due to relatively quick exhaustion of reserve mechanisms and enzymatic storage.

Saying this, in extreme conditions there is no clear correlation between synthesis and the secretion of CL, ACTH, and corticosteroids. The adrenal gland gets activated less than the increase in CL and ACTH. This suggest that the central links of the HHA system possess more liability and reactivity compared to the distal (target organs) parts.

Another link in parthenogenesis of stress is the increasing role of the para-hypophyseal pathway regulation of adrenal cortex activity, based on direct CNS and humoral influences, due to accumulation of biologically active substances.

The synthesis and secretion of corticosteroids in extreme conditions varies not only quantitatively but also qualitatively. More often, in steroidogenesis, there was noted a shift towards cortisol and aldosterone secretion with absolute hydrocortisone insufficiency.

Changes in the transport of hormones is often disrupted due to the inhibition of the corticosteroids, coupling with corticosteroid binding globulin (aka storage). This facilitates the rapid accumulation of active steroids in the bloodstream as well as easing its cell's entrance.

Next comes another typical syndrome of stress in extreme conditions: corticosteroids insufficiency due to resources exhaustion. According to a mechanism of its development it has several forms. First is adrenal glandular form, associated with impaired synthesis and secretion of steroids. Second one is pre-adrenal, seen in extremely sick patients, due to the increased binding of steroids with plasma proteins and erythrocytes, or the inability of tissues to metabolise/eliminate it. The result is the same: a lack of active steroids. The third form is the so-called: 'relative steroids deficiency', based on a mismatch between the sharply elevated corticosteroid's demand on the body and the ability HHA system to meet this request. The fourth, and the most commonly encountered one is combined form, which can have elements of each of the above.

Stress reactions as a response to extreme conditions have some specific presentations. First of all, the severity of the HHA system impairment, as well as various extreme conditions will be manifested by different shifts in metabolism and organ dysfunction. For instance, in the case of traumatic shock and burns, the high activity of the HHA system is formed faster and remains longer than in haemorrhagic shock and anaphylaxis. Clear differences are found in the hormone spectrum, secreted by a cortical layer of the adrenal gland. In traumatic shock, a shift of steroidogenesis towards a secretion of corticosterone happens rapidly, while in burns, this is noted only in the terminal period. Something similar can be seen in haemorrhagic shock. Therefore, different forms of shock will trigger direct changers in the synthesis and secretion of different quantities of hydrocortisone, aldosterone, pregnenolone, progesterone, 17-a-oxyprogesterone and 11-desocortisone.

Certain characteristics have been noted in the transportation mechanism. In shock, there was observed a significant accumulation of corticosteroid content in red blood cells. Another important feature: in deep hypothermia, there is increase in reserving steroids due to stimulating binding steroids to protein-transporters. And the elimination of hormones can be very dependent on the type of a shock.

The role of stress response under extreme conditions is largely determined on the effect of catecholamines and corticosteroids. The basic function of these hormones is urgent energy production, as well as mobilisation of all functional reserves of a body. Current studies suggest that not only mobilisation of energy and resources occurs in stress but also its redistribution in the direction of provision of the system operating with maximum load (where it is most needed).

Thus, it's firmly established that stress-response plays a crucial role in adaptation and compensatory processes in extreme conditions. Originally weakened functions of the HHA system leads to a rapid deterioration in a patient's condition and to bad outcomes, while timely ACTH and steroid administration for treatment and prevention gives positive results.

However, with excessive long stress duration under extreme conditions, there is evidence of negative consequences. An excess of catecholamines causes the centralisation of blood flow, and microcirculation disorders, dangerous for parenchymatous organs. Tissue hypoxia, metabolic acidosis, and the activation of lipid peroxidation leads to damage of the cell membranes. A high concentration of glucocorticoids in severe stress stimulates catabolic processes (in protein synthesis), disrupting the immune system, resulting in decay and migration of lymphatic cells. Thus, the phenomenon of hyper-compensation in stress responses leads to the fact that this reaction can easily turn from physiological to pathological disorder, contributing to the severity of the homeostasis problem.

In the psychiatric literature the problem of acute stress development is well established, while very little known about chronic stress disorder, so-called 'endemic stress'. The latter is understood as the state of mental illness caused by the constant negative influence different social pressures (family, career conflicts, impacts of noise, vibration, environmental pollution, threat of war, etc.). Endemic stress spreads easily to large groups of people and entire nations. The consequences of this more often remain hidden for a long period of time (left in subclinical level). It's usually judged retrospectively, after clinical presentation for some other syndrome or disease (mental or psychosomatic). Detection of such pathology can be provoked by acute stress. Development of an acute stress on a top of chronic one can aggravate the situation.

At present, another stress is distinguished: catastrophic – a condition caused by super strong cataclysms (wars, natural disasters), affecting populations of people.

The stress in modern psychiatry is also associated with increased usage of psychotic, neuroleptic and antistress medications (especially benzodiazepines), drugs interactions and related to these medical iatrogenic complications.

2.3. Adaptation Syndrome

This is a set of non-specific changers, occurring in the animal and human body under the influence of any pathogenic irritant. The term was proposed by Selier in 1936. According to Selier, an adaptation syndrome is a clinical manifestation of stress-response, always under adverse conditions for a body.

Selier classifies between a general or 'generalised adaptation' syndrome, with shock as its clinical feature, and local adaptation syndrome, developing in a form of inflammation. The generalised syndrome is so named because it affects the whole organism, and an 'adaptation' means the healing process follows.

The general adaptation syndrome has a few stages. In the beginning when there is a danger to homeostasis, there is seen the mobilisation of the body's defence forces, so there is a stage of anxiety (alarm – a call for mobilisation). The second phase is the restoration of disrupted equilibrium and the transition to a stage of resistance, when the body becomes insensitive not only to this irritant but to other pathogenic factors too (cross-resistance). If a body does not fully adapt to a pathogen's influence, then the third stage develops: the exhaustion stage. The cell/organism death might occur in the anxiety or exhaustion stages.

One of the indicators of either stage of adaptation syndrome may be a fluctuation of the basic metabolism. In the first stage there is the prevalence of catabolic reactions (dissimilation stage), and at the stage of resistance anabolism dominates (assimilation stage). In animal studies (rats) as they are constantly growing, the stages of adaptation syndrome may be distinguished by daily (electronically) monitored weight changes.

The most significant changes in a body in general adaptation syndrome are: adrenal cortical hypertrophy, atrophy of the thymus-lymphatic system, and bleeding ulcers in a stomach or duodenum. This was noted even before Selier's work. Selier was determined to find out the causes of an adaptation syndrome and distinguish its nature. He successfully solved part of this very different task. It is established that many changes in general adaptation syndrome are due to increased activity of the anterior part of pituitary gland, ACTH secretion and adrenal cortex stimulation. However, the stress-response theory reflects only the phase nature of nosogenous trophic disorders and therefore not used in personalised medicine.

Many researchers have shown that the reaction of the anterior pituitary gland and adrenal cortex occur very rapidly (minutes and even seconds) and then in turn, it depends on both the hypothalamus (producing releasing factors) and the acid-base state (osmotic pressure).

Under conditions of excessive or prolonged exposure to damaging factors, there may occur significant deviations of the body's constants, sometimes beyond tolerable limits, which leads to the disruption of normal physiological functions and the development of pathological processes.

In addition to constant maintenance through an adaptation, various functions of the body can be rearranged to ensure an organism's survival against physical, emotional and other pressures.

2.4 Generic Categories of Energy-Structural Status

The reliability of a treatment safety determines the level of energy-structural status (ESS) of body cell mass (BCM), which can ensure the implementation of targeted genetic programming for numerous extracellular molecular signals. As genetic programs change in accordance with its replacements, the level of BCM energy-structural activity also changes. The energy production shifts become effective when its intensity is compared to the level of proper basic metabolism (PBM), which requires for a patient's BCM to be able to carry out genetic programs, i.e., manifest its biological potential.

Energy is truly considered the foundation of biological evolution. However, for stable energy production equilibrium, it is necessary that the utilisation of energy substrates, a transfer of electrons of C4 and H+ organic molecules is carried out on O2. Its biological significance is clearly demonstrated by comparison with the intensity of O2 transport with its proper value, reflecting the energy-protective ability of the patient's functional state, i.e., its energy-protective potential.

A comparison of the values of the current energy-protective status and biological potential with corresponding level of metabolism (1202 kcal/m^2) and VO2 (600 ml/min × m^2) provides reliable survival in critical patients. It also allows us to establish the nature of energy biotics (eu-, hyper-, hypo-, patho-) relating to energy protection (eu-, hyper-, hypo-, patho-), calculated in percentage.

The relationship of these parameters is presented below in Table 2.1

Categories of energy structural status	Biological Potential	Energy-Protective Potential
Eu-energy structural status	Eu-energybiothy	Eu-energy protectivity
Dysfunction of energy-structural status	Hypo-energybiothy Eu-energybiothy Normo-energybiothy Hyper-energybiothy	Eu-energy protectivity Hypo-energy protectivity Hyper-energy protectivity Eu-energy protectivity
Damage of energy-structural status	Hypo-energybiothy Hyper-energybiothy	Hypo-energy protectivity Hyper-energy protectivity
Deficit of energy-structural status	Patho-energybiothy	Hyper-energy protectivity Eu-energy protectivity Hypo-energy protectivity
Failure of energy-structural status	Patho-energybiothy	Patho- energy protectivity

Table 2.1 Volumetric categories of energy- structural status

It follows from the presented data that ESS dysfunction can occur with either BP or EPP disturbances. The main damaging ESS factor is hypo-energybiothy, caused by factors responsible for hypo- or hyper- energy protectivity. Even with the energy still delivered or energy resources still preserved in case of patho-biothy (unable to energy distributing, for a sample) a deficiency in ESS will still occur. Alteration of BCM is characterised by Pathoenergy-biothy and patho-energy protectivity, and failure of ESS. The safety of medical care in this case becomes doubtful. It's especially true in surgery and anaesthesia. The use of energy-protective technologies to ensure the elimination of ESS shifts will restore the safety of a therapy. Satisfaction of energy structural requirements is a guarantee of the elimination of life dangers. Here comes numerical perinosal vitalism, which states the discrepancy between the actual values of ESS and energy structural needs, expressed in a percentage. The prevalence of a level of credibility in ESS and ignorance of BCM needs are dangerous and ineffective scenarios which will bring no result or even can be harmful.

The exact boundaries of safe and hazardous energy biothy values, linked to ESS categories, are shown in Table 2.2

Category of energy-structural status	Energobiothy (kcal/m²)		Vitalism (%)	
	Energy-structural activity	Metabolism needs	Reserve	Deficit
Eu-energy protective	990-1375	1125-1185	3,7-8,6	-
Dysfunction: hypoergic hyperergic	745-983 1376-1734	895-1082 1286-1472	3,3	14,2 6,6-18,3
Damage: hypoergic hyperergic	661-980 1735-2700	845-1036 1473-1975	-	1,3-19,6 18,4-39,2
Deficit: hypoergic hyperergic	353-660 2710-3900	660-844 1976-2910	-	37,1-19,6 39,2-58,7
Failure	212-352	565-659	-	48,6-37,2

Table 2.2 Vitalism of energy-structural categories

The presented data state that the safe level of energy biothy can only fluctuate between 15% – 19%. It proves that securing energy-protective health care is a complex task, and its resolution requires high accuracy in maintaining the safety of each single component of energy protective cellular vitality.

ESS dysfunction with minimal hypoenergobiothy characterises by a tendency to eu-energy biothy, which without energy protection leads to vitalism deficiency, and has around 14% risk of energy structural disorders.

Hyperergic dysfunction is fraught with a constant deficit of vitalism at least of 18.3% in value.

Deeper vitality insufficiency contributes to a risk of damage, inadequacy and insolvency of energy structural status (ESS).

At the same time hyperergic mechanisms of disorders considered more vitality damaging than hypoergic ones...

2.5 Energy Protective Reserves

Reliability of eubiothy is determined by such state of a body, which can speed up biocycle 'energy production – energy delivery' to cover for the increasing demand of the microcirculatory-mitochondrial (MM) complex. The quantum biocycle (QB) is the main factor of reliability ESS as it is a measure reflecting how energy biothy depends on adequate oxygen and energy substrates delivery.

Heart rate (HR) determines the duration of quantum and therefore can be used to compare submaximal HR with its actual figure, defining myocardial reserve. In case of total cardiac ischaemia, coronary perfusion is dropped, and myocardial reserve is lost. ESS oxygen transport reserve forms that possible increase in oxygen delivery (DO2), the later can provide the required current energy biothy.

The value of reserve biothy is larger than the current biothy one. The calculation of ESS reserve is convenient to perform with the use of the factor of Qx (belongs to deep oxygen transport status) – the factor of compensation O2 deficiency. If Qx > 1 there is ESS deficiency, while a value (Qx-!) corresponds to the relative deficit of energy protection. The ESS is formed by an MM complex, which produces more energy than required at the time. A calculation is carried out using current arterio-venous difference O2 and comparing it with the one needed for energy biothy activities (index Cx from deep oxygen status equation). The failure of ESS of an MM nature manifests by prevalence of Cx over the current oxygen extraction. This condition will define a deficit of energy.

2.6 Destabilisation of Autoregulation

Stabilisation of the energy structural status of BCM requires constant cytoarchitectonics to support cytoskeleton and intracellular fluid. The ions equilibrium determines the osmotic pressure, which makes up transmembrane transport of ions and water. Thus, changing can cause deformation of a BCM and compromise the mechanisms of energy structural status. The relationship between main components of the oxygen equation and of plasma osmolality allows us to estimate the level of energy-osmolar stability. The latter is preserved if real osmotic pressure does not fluctuate more than +/- 3% from a value corresponding to the need for energy biothy and energy protection. Destabilisation has hyper- or hypo-osmotic character in nature. If autoregulation responds to osmolar disturbances in milliseconds, the haemodynamics takes 2+ minutes to respond. Therefore, even with the minimum changers in BP and a ratio of systolic/diastolic blood pressures (SAD/DAP) are very dangerous. The ratio SAD/DAP reflects the relationship of blood flow and its oxygen distributions to organs. The figure of a ratio (0.599–0.636) corresponds with achieved stabilisation of hemodynamic autoregulation energy-structural status. The rise above the upper level of the interval reveals diastolic destabilisation, a drop below the lower limit, a systolic one. The representation in percentage emphasises the simplicity of the interpretation

2.7. Energy-Structural Status Properties

The condition of ESS is reflected by its properties: adaptability and destructiveness, stability and instability, adequacy and inadequacy.

Adaptability refers to the ability of BCM to accommodate additional energy-structural support in response to new genetic programs requests. Destructive ESS occurs if the intensity of biologic potential (BP) is insufficient, the energy structural support fails, and genetic programs suffer.

The ability to increase ESS in response to body cell mass activation reveals the comparison between current provision of energy and the level of it required. If the latter kept below the level of real BP, that the increase in energy biothy of BCM will be labile.

If ESS is unable to satisfy the energy demand of BCM it is evidenced by the instability of ESS.

The adequacy of ESS of BCM equals a sum of adaptability and stability, while if destruction and instability manifest, we declare inadequate ESS.

2.8. Vitalism Space

The eu-biothy is declared if for the current needs of energy-protective status a patient uses from 130-160 ml/min x m2 of oxygen. Moreover, in order not to deplete the reserves it is recommended to implement energy substrates with a total potential 1040–1160 kcal/m2 daily.

An audit of ESS shows that a vitalism of ESS defines eubiotic space quite precisely (Table 2.3)

Parameters of Vitalism	Limitation of eubiothy, %
Reserve of vitalism	3,7-18,6
Myocardial reserve	49-55
Oxygen transport reserve	12 - ÷
Oxygen delivery insufficiency	÷ - 6
Microcirculatory-mitochondrial reserve	4,7 - ÷
Microcirculatory-mitochondrial insufficiency	÷ - 12,2

Table 2.3 Audit of vitalism and ESS reserve

The eu-biothy stability mostly depends on myocardial reserve (49–55%), this excludes the risk of hyperenergy protection, which leads to the failure of oxygen transport.

This also applies to the MM complex, where hyper energy-biothy has exhausted its reserves.

Therefore, vitalism makes it possible to personify the maintenance eubiothy, the boundaries of which are defined not by current but predicted values. The latter ones are determined by the nature of intensified genetic programs.

CHAPTER 3

Nozogenous Euvitalism

3.1 Autoregulation of Vitalism

The main purpose of it to preserve the self-regulating bio-sustainability of an organism, the key of this is the constancy of the ratio mass, volume and the surface of its cells. The lack of cell deformities serves as reliable evidence of the cell's optimal function.

The conjugation of mass-volume-surface provides water. Due to water's unique properties, all cell compartments are filled with water and all BP reactions take place there. The energy-structural conjugate factor of water is the polarity of its molecules, despite its el neutral charge.

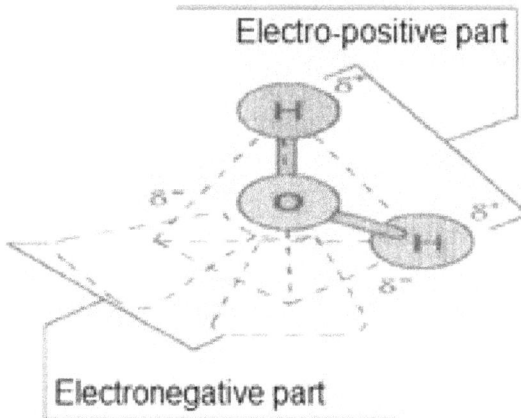

Fig.3.1 Water polarity

The matter is that in a molecule of water, the O2 nucleus attracts the nucleus of surrounded hydrogen atoms. This polarisation is responsible for forming a hydrogen bond, which is 20 times weaker than covalent one. However, it does not stop water molecules combining into a spatial lattice.

Their cohesive nature explains properties of water such as high surface tension, high specific heat capacity and optimal evaporation. At body temperature, around 15% of water molecules retain this conjugating ability. Each molecule connects to four neighbouring molecules and forms 'blistering clusters' (short-lived aggregates). They also group around other ions and molecules. Therefore, the compounds, participating in polarisation, will dissolve in water, earning a name: hydrophilic. Non-polar molecules destroy H-water bonds, and so do not dissolve in water – i.e. they are hydrophobic. Such hydrophobic groups, surrounded by water, tend to get closer, making the water structure stronger, and less disturbed.

3.2 Water-Electrolite Interaction

Energy conjugation determines the total amount of water in a body, which depends on sex, age, amount of subcutaneous fat, etc. On average it's about 60% of body weight. The intracellular compartment makes up 40% of it, the extracellular 20%, while 5 % is intravascular water, and 15% interstitial one. These compartments are functionally united and interact closely with the intracellular space. There is also distinguished the third water space: transcellular (CSF, intestinal mucus, serous fluids). This can be ignored in normal circumstances.

The composition of water compartments shown on the diagram of Gamble.

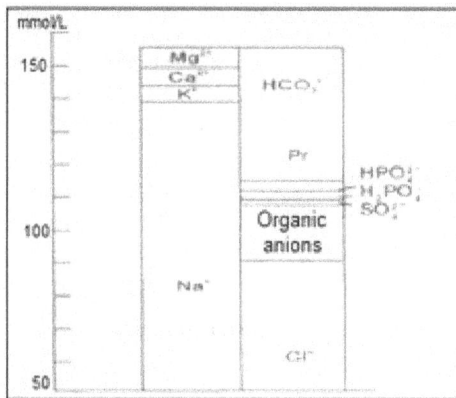

Pict 3.2 Extracellular concentration of ions

The sum of cations and anions ensures the neutral charge. In extracellular fluid Na+ dominates, and in the intracellular potassium cations, and jointly this determines osmolality. As to anions: Cl- and HCO3 belong to extracellular space, while proteins and phosphates predominate in the intracellular compartment.

Plasma interstitial fluid has a small amount of a protein. According to Donnan's equilibrium, anions are found on the protein-free side, while cations will be on the intracapillary side, the part containing proteins. Water compartments are always on motion. Its dynamic constancy is secured by complex physical, chemical, neuro-humoral and metabolic processes. The water–electrolyte balance is only maintained if the administration of liquid corresponds with its release.

Type of fluid	Amount, ml
Drinks	1000-1500
Food	700
Endogenic water	300
Overall:	2000-2500

Table 3.3 Fluid requirement daily

In a healthy person, fluid intake is regulated by thirst. Water in different quantities is contained in food. Endogenous water comes from nutrient oxidation with hydrogen. The amount of urine depends on the kidney's filtration and on the body's level of hydration.

Evaporating through the skin or lungs, water does not contain electrolytes and counts as insensible losses. Any pathological processes that increase an evaporation or tachypnoea will result in the loss of several litres of water. For instance, an increase in body temperature by one degree Celsius increases the loss of water by 300–400 ml (see a Table 3.2).

Water loses	Amount, ml
With urine	1000-1500
With sweat	500
Through lungs	400
With feces	100
Overall:	2000-2500

Table 3.4 Insensible loses

In contrast to evaporated water, electrolytes are lost through sweat. Releasing water with perspiration, during vapour anaesthesia, is called external water exchange. The exchange of large volume of water at the capillary, lymphatic, venous, intestinal systems or distributed through the kidney or plasma is called internal water exchange.

Daily gastroenteric water circulation is presented in Table 3.3

Body fluids	Amount, ml
Saliva	1000-1500
Gastric juice	2500
Pancreatic juice	1000
Bile	750-1000
Intestinal juice	3300
Overall:	8000-9000

Table 3.5 Daily gastric water exchange

Water compartments are in dynamic equilibrium. If the normal exchange is disturbed, control mechanisms are immediately activated to restore the balance. Changes in the volume of water spaces inevitably leads to electrolyte impairment.

The infusion of crystalloids affects only the extracellular water compartment. However, the latter, constantly changing, ensures the constancy of intracellular space.

As clinicians we can only correct extracellular sector.

See Figure 3.3: Relationship of intracellular water and Na+

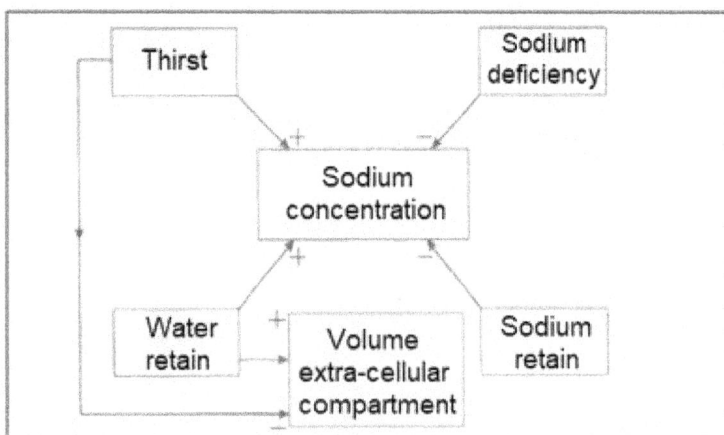

Fig. 3.6 Relationship of electrolytes balance

The osmotic pressure in balance fluid compartments is the same. The accumulation of K+ intracellular and Na + in blood is the result of cell metabolism. It is called a sodium-potassium pump.

3.3 Regulation of Electrolyte Balance

It is impossible to allocate an organ or a system which does not participate in the water–electrolyte exchange.

When discussing the usual scheme of water–electrolyte equilibrium used in clinical practice, it is important to outline the following principles.

There is a volemic regulation, triggered by low pressure baroreceptors (volume receptors) mainly in the atria, carotids, and possibly other (even interstitial) vessels. It is likely that volume receptors of different zones will have a different response.

An increase or drop in blood volume causes the stimulation of hypothalamic centres, pituitary, adrenal glands, leading to a delay or increase of fluids filtering through a kidney.

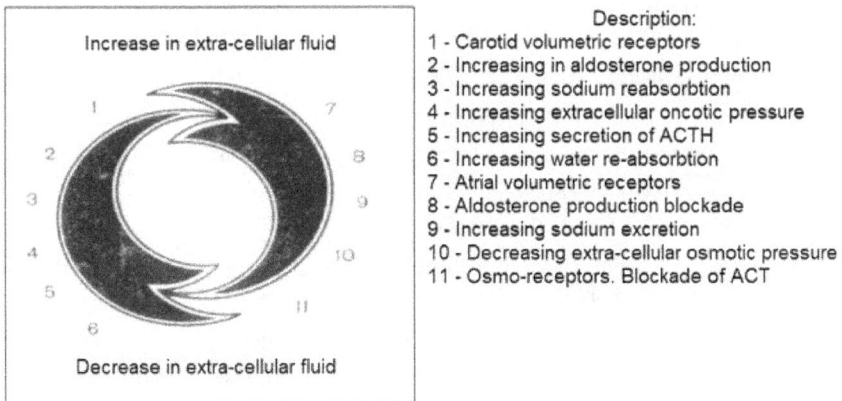

Description:
1 - Carotid volumetric receptors
2 - Increasing in aldosterone production
3 - Increasing sodium reabsorbtion
4 - Increasing extracellular oncotic pressure
5 - Increasing secretion of ACTH
6 - Increasing water re-absorbtion
7 - Atrial volumetric receptors
8 - Aldosterone production blockade
9 - Increasing sodium excretion
10 - Decreasing extra-cellular osmotic pressure
11 - Osmo-receptors. Blockade of ACT

Increase in extra-cellular fluid

Decrease in extra-cellular fluid

Fig 3.7 Volume regulation of extracellular fluid (by Zilber A.P)

Osmoregulation is even more sensitive if it started on the level of osmoreceptors of the extracellular compartment. These receptors are possibly present in all organs, but indefinitely seen inside neurons in the subthalamic area (looking like neurons with vacuoles). Those one has strictly defined an osmolality level (300 mmol). Activated by hypotonic or hypertonic extracellular fluids, shifts into neurons of these osmoreceptors acts similar as the stimulated volume receptors in vessels.

The combination of volume and osmoreceptors regulate activity of the antidiuretic hormone (ADH) and aldosterone secretions. As a result, the blood volume and osmolality of extra and intracellular spaces remained unchanged.

To summarise, the water–electrolyte equilibrium is maintained mainly by changing in a water reabsorption and sodium exchange (ADH and aldosterone).

The changes in a body, arising nosogenically is a compensatory adaptive reaction aimed at restoring the stability of oxygen transport and maintaining homeostasis (mass, volume and surface of cells). The sympatho-adrenal system, kallikrein-kinin cascade, histamine and serotonin release, react sensitively to slightest changes in the neurochemical pathway, taking an active part in perinozal maintenance of oxygen regime.

3.4 Conjugation of Autoregulation Components

The relationship between metabolism and oxygen transport is determined by the oxygen regime reading (ORR), which is the ratio of oxygen delivery to oxygen consumption. The optimal compliance of these vital parameters serves as a criterion for balancing the ORR (from 3 to 3.5) and energy-protective state. When oxygen consumption dominates the oxygen delivery it indicates ORR under pressure (< 2.9). Excess O2 delivery should be considered as excess ORR (ORR > 3.6).

The characteristics of the functioning vitalism's mechanisms will be incomplete if we do not take into consideration the intensity of O2 extraction by tissues. Therefore, in the case of a dangerous microcirculatory-mitochondrial deficiency syndrome (MMDS) it is necessary to estimate the condition of vascular-tissue O2 transport. This important parameter can be calculated by estimation of the quantitative determination of O2, extracted and utilised by tissues from each unit of red blood cells (mlO2/L).

The complete characteristics of the circulatory system can be characterised better if, together with the determination of the final result, we would record preterm adaptive parameters. For instance, preterm adaptive haemodynamic figure is determined by the value of the cardiac index (minute volume of circulation). The latter best reflects the ability of blood circulation as part of the oxygen delivery system in meeting BCM metabolic demands.

However, the homogeneous system-forming factor can act in many ways: through a heart, microcirculation, vessels of blood collection and its distribution vessels. Taking their predetermined adaptive value results into account, the evaluation of the functional circulatory system should measure absolute figure of CI and its integral indexes, haemodynamics, microcirculation, flood flow and its distribution.

According to the circulatory system's ability to provide the bioenergy body's needs, its general condition is assessed. For this purpose, first it is necessary to distinguish to which of the seven main types of haemodynamics the calculated CI value belongs.

The absolute of cardiovascular system performance is then compared with the level of the effective CI. This makes it possible to objectively estimate the type of haemodynamics (hyper-, hypo-, or normo-) and obtain characteristics of blood flow.

The comprehensive assessment of haemodynamics is refined by determining preliminary cardiac function. It is considered that the best estimation of cardiac function gives the fraction of CI, equal to the ratio of the cardiac stroke volume to end-diastolic blood volume.

Although dynamic monitoring of changes in stroke volume in the process of energy-bio correction is not always possible technically, this parameter also only characterises the cardiac function in this single time moment. The other independent parameters such as the maximum speed of the shrinking of myocytes (Vmax) and the speed of BP increase in left ventricle (dP/Dtmax). It is also not that useful in cardiac function monitoring as it is difficult to implement in practice. Overall, at the present level of knowledge, it's theoretically impossible to find indexes, which will clearly determine the effect of cardiomyocytes activation, and also factors affecting myocardium.

The implementation of phase analysis allows us to abandon the 'complicated for practical use' of cardiac indexes but has its limitations in estimating cardiac function in terms of its bioenergy profile. Therefore, the fundamental factor of cardio-dynamic evaluation should be considered the matching of myocardial oxygen consumption to the stroke volume of a heart. Its evaluation during the period of bioenergetic correction should be continuous, which is difficult to do in practice.

However, it is the product of average BP multiplied by HR, that is used more commonly to estimate the O2 consumption by a myocardium. Its correlation with cardiac performance makes it possible to assess precisely the bioenergetic efficiency of myocardium as it determines the volume of blood in each contraction per unit of oxygen consumed.

The correlation analysis established a highly reliable association of coefficient cardio dynamics (CCD) with the cardiac output fraction: the independent correlation coefficient was 0.68.

The method of nonlinear evaluation helped to establish the nature of the CCD connection with systolic (Ds) and diastolic diameter (Dd), end diastolic (ENDV) and minute blood volume (MBV).

$$\text{CCD} = 1.06 - 1.1\text{Dd/Ds} + 0.00002\ \text{Dd} \times \text{ENDV} \times \text{MBV} - 0.05 \times \text{Dd}$$
$$(\text{ml/HR} \times \text{kPa})$$

This equation, with high degree of reliability ($f= 5.33$ at $F0.95= 1.96$) indicates that CCD, taken integrally, expresses the complex relationship common laws of haemodynamics.

The current assessment is made by comparison of the actual value and corresponding vitalism-saving level of CCD.

If the ratio is 0.9–1.1, it corresponds with normodynamy,

O.8 and <: hypo cardiodynamy,

1.2 and >: hyper cardiodynamy.

monitoring of the ratio actual and vitalism- supplying KKD allows us to identify a decrease in myocardial vitality before the manifestation of clinical signs of cardiac weakness.

Estimation of the pre-term adaptive results of the haemodynamics should be made according to the average arterial BP. The latter plays a leading role in regulating the degree of Hb deoxygenation. Some specific information of different levels of the stimulation of the metabolism is transmitted by restored Hb. It helps us estimate of blood distribution to tissues dependably on the coefficient of vasodilatation (CV).

If the CV is equal 0.9–1.1 it fixates the normo-vasotony. With VC 0.8 and less, it indicates hypo-vasotony, while at 1.2 or more we consider hyper-vasotony. When accessing the vasomotor state of the vessels, it should always be remembered that the obtained information reflects the duration of contact between blood and capillaries, that is necessary for effective oxygen utilisation.

The preliminary result of vessel function is best to characterise the main indicator of volume status – the stress blood volume (SBV), which depends on venous return and the stroke volume. At the same time it should be constantly taken into account that only SBV defines an ability to provide the system transport O2 at the level necessary for vitalism maintenance and adequate vasomotion.

Microcirculation disorders play a leading role in the pathogenesis of vitality impairment. In order to diagnose it, a conjunctival bio- microscopy is performed, and a check of the arterial-venous ratio of protein concentration and haematocrit (HT). The results, although obtained from a single vascular area, can be projected to the whole body. Probably that is why the clinical assessment of the microcirculation state remains speculative.

The ratio of areas of capillary filtration and reabsorption determines the ratio of interstitial fluid volume to the volume of circulating plasma. This ratio, as a generalised indicator of transcapillary exchange in a body, summarises filtration and reabsorption flows, which depends on arterial inflow into capillaries, outflow of venous blood from it, pre- and post-capillary resistance, properties of interstitium and vascular permeability.

A correlation analysis shown that the index transcapillary exchange (ITE) reversibly proportional to the shape of a vessel (r = -0.544) and general cojectival index (r = -0.522) Thus ITE really reflects the process of histohaematic permeability caused by the vascular mechanism, for example the ratio pre- and postcapillary resistance.

Microcirculation is assessed by comparing the actual value with the corresponding vitalism parameter.

If the ratio is 0.9 to 1.1, then microcirculation is unaltered. The ratio 0.8 and < and 1.2 and > allow to identify a hypo- and hypervelocity state of microcirculation. It gives you the understanding that only in this way you can receive enough flow to facilitate blood exchange in a level, unlimited of homeostasis of mass-volume-surface BCM.

One of factors limiting the restoration of vitalism is osmolality disturbances. Osmolality is involved in all mass exchange, which is performed at the level of capillaries, uniting tissues with body fluid compartments.

The vitalism supported osmolality (VSPO) level, in a calculation in which the violation of the individual system O2 transport and its consumption (VO2) should be excluded, can be equal to:

$$\mathbf{VSPO = 332 - 0.26 \times dDO2 - 137 \times Vo2, \ mosm/l}$$

Evaluation of the osmotic state of vitality preservation is carried out by a comparison of the value of VSPO from the actual VSPO, or with the one, calculated by the equation:

VSPO = 1.86C Na + C x(glucose) + C (UN) + 5, mosm/l. where

C Na – concentration of Na
C K – concentration of K
C (UN) – urea nitrogen level

Using the deviation figure of VSPO from the actual VSPO, or we call it the discriminant of critical osmolality, we can assess the osmolality shifts and test its ability to limit body's inner energybiotic need. In doing so we will increase the energy-saving role of anosmia management.

3.5 Natriyuretic Regulation

Blood volume, BP, sodium and diuresis are regulated by natriuretic peptides. They also affect vessel relaxation and remodelling, myocardial hypertrophy and fibrosis, lipid metabolism and the growth of long tubular bones. In mammals the members of its family are atrial natriuretic peptide (ANP), B-natriuretic peptide (BNP), C- natriuretic peptide (CNP) and possibly osteocrin/muslin. Determining the level of the above peptides in patient's plasma is used for many disease diagnostics, and some peptides even have therapeutic use.

The role of natriuretic peptides in nosogenous activation is multifaceted. ANP expands arterial vessels, and stimulates Na excretion through kidneys, regulating by this PB and salt-water homeostasis. The physiological effect of BNP includes diuresis regulation, suppression and activation of many cytokines, endothelins, sympathetic system and the renin-angiotensin-aldosterone mechanism.

ANP, through specific receptors, causes relaxation of smooth muscles in vessels, vasodilatation and a decrease in BP. In addition, ANP inhibits the release of aldosterone, angiotensin 2, endothelin, renin and vasopressin. In the kidney, ANP has three functions. Firstly, it increases the glomerular filtration rate by rising pressure in glomerular capillaries, causing differentiated dilatation afferent arterioles and contraction of efferent vessels. Secondly, ANP decreases reabsorption Na and water in different parts of nephrons through guanyl-monophosphate (GMP) dependable modulation of sodium channels and its transporters. For example, in proximal tubules, ANP inhibits the Na and water reabsorption, stimulated by angiotensin 1. In distal convoluted tubules, ANP reduces Na reabsorption by inhibiting amyloid-sensitive cation channels. Thirdly, ANP decreases secretion of renin by the juxtaglomerular apparatus through the GMP-protein kinase molecular pathway.

These processes all together decrease natriuresis, diuresis and renin secretion. This explains why in patients with cardiac and chronic kidney failures and severe HTN detected high level ANP.

ANP continuously regulates the transcapillary fluid balance, increasing capillary hydraulic conductivity and endothelium permeability to macromolecules, similar to albumin. The hypovolemic mechanism of ANP action is still unclear. However, it is revealed that an ANP-caused drop in intravascular volume does not require its natriuretic or diuretic properties of a hormone, as it precedes urination and occurs in nephrectomised animals.

ANP and BNP through the receptors of smooth muscles cause a relaxation of preliminary contracted circular muscles of aorta, that is most likely the mechanism of the elimination of acute hypertension (non-treated one).

As to CNP, it's released in response to damage of vascular endothelium and inhibits the proliferation of vascular smooth cells, as well as migration cells of coronary vessels. This reaction is induced by low density oxidised lipoproteins. CNP, once released from cardiac endothelium, plays a cardioprotective role by inhibiting ischemic reperfusion to a damaged heart.

Balanced energy conjugation of BCM is characterised by optimal free energy production and effective level of all biosynthetic processes, which is the aim of euvitalism. It's the guarantee of self-preservation and self-regulation, the harmony of energy-structural interactions. It is all secured by energy-protective mechanisms and satisfies physiological needs of an organism in water, electrolytes and energy components.

3.6 Restoration Of Structural Elements

BCM consistency in physiological conditions is ensured by the mechanism of self-renewal (regeneration) of specialised cell populations and cell organelles. And in the case of cell damage, by the processes of reparative regeneration.

Overall, there are about 200 different phenotypes of specialised cells in adult tissues (Albert et al. 1994).

At tissue level, the following concepts are distinguished: clone/cell population, differon, vascular cell matrix, tissues.

A clone of cells is the population of descendants of one stem cell differentiated in one direction to perform a phenotypically single-profile cell function (neurons, astrocytes, oligodendrocytes, hepatocytes, cardiomyocytes). Differon is a set of single profile clone cells integrated into tissues to perform tissue specific functions. For example, gliotic nervous tissue differon is a population of specialised astrocytes and oligodendrocytes which perform CNS orders for impulses delivery. The keratinolytic skin epidermis the differon makes up the population of single-profile epithelial cells, forming a multilayer cover of the keratinised epithelium with defined protective function.

The vascular tissue cells matrix is a set of blood and lymphatic vessels, as well as molecular-fibre extracellular bed, which together with a population of specialised cells provides fluid/electrolytes/oxygen exchange.

An organ is a morpho functioning system of vessels, ducts, basal membranes, molecular fibre extracellular matrix, plus specialised cells integrated into specific microstructures of certain architecture (nephrons, alveolar acinuses, etc.) to perform the specific functions of an organ.

The process of restoration of structural elements of tissues and organs instead of the lost ones is carried out on four levels: molecular-membranous, subcellular, cellular and tissue-organic. Regeneration can be physiological and reparative; and a regeneration disturbance is commonly called disregenerarion. Dependably on molecular features reparative regeneration divides into regeneration on damaged specialised cells and on damaged organs, the latter is more complex.

Physiological self-renewal is a continuous process of updating the population of specialised cells that are able to perform genetically programmed specific functions. In organs, a regeneration is morphologically observed in four processes: rhythmic mitotic division of stem and fixed cambial cells, elimination of appeared defective cells by apoptosis, differentiation of new cell generation and their migration to the target regions of an organ as well as death cells that have exhausted their life cycle. The process of cell reproduction observed on evidence of mitosis was called proliferation of cells, while the phenomenon of increasing the number of certain cells was defined as cells hyperplasia.

Cells transition from functioning to division; differentiation of new cells populations or apoptosis is dictated by activated gene cascade and triggered by molecular signals from the extracellular syncytium. Gene-initiators of cell multiplications are growth-regulating genes (proto-oncogenes), genes-suppressors: these are tackling uncontrolled cells proliferation (antioncogenes), genes, that provide a cell's genome stability and adequate transmission of information during mitosis.

Other genes include genes-regulators of cell differentiation, pro- and antiapoptotic genes. Like all components of genome, these genes perform their specific functions only under oxygen-protective conditions.

CHAPTER 4

Energy Cellular Resuscitation in Vitalism Impairment

4.1 General Conditions Leading to Disorders

Pathomorphological characteristics of energy-structural damages are closely linked to alteration of structure of cellular membrane (shape), cell volume, its organelles quantity and quality. Structural equilibrium is the condition in which maintenance is most energy-consuming process of BCM. Sparing more free energy for cell structure and surface stabilisation, especially in case of Hyperenergy biothy, inevitably leads to reduction of energy supply of self-recovery mechanisms and may cause cells apoptosis. Salt-water disbalance in case of impaired energy-structural conjugation impairment depends on mismatch between endotoxicosis and the amplitude of genome response.

Primary damage to the transport across membranes occurs under the actions of bacterial toxins, viral proteins, lysis of fragments of the complementary system, lymphocytes perforins, membrane poisons, toxic medicines and sudden changes in osmolality.

Below is list of adverse consequences of impaired transports of the cell membrane:

- Violation of transmembrane exchange, active transport of ions, water through plasma synthitium disrupts the intracellular water balance, leading to two stages of hyperhydration: either swelling or cell oedema cell swelling with excessive hydration of cytosol and karyoplasm with a simultaneous increase in the volume of cytoplasm and cell nucleus. Cell oedema – excessive hydration and vacuolisation of tubular components of endoplasm with less significant increase in the cell volume.

- Decreased activity of Na-K ATP, which impairs active transport ions, contributing to swelling. Further extra hydration of cytosol may result in cell decomposition.
- Disturbance of Na-K ATP plasmatic membrane of ependymocytes vascular junctions of a brain, causing disruption in water-proteins transport and breakage CSF barrier.
- In conditions when single-valence ions transport is disrupted, there is an increase in passive transport potassium out of cells, causing Na and water flow changes and penetration of the cell with hydrogen and glucose. This results in extracellular hyperkalaemia and intracellular acidosis with its consequences.
- Dysfunction of calcium-dependent ATP in energy deficiency leads to accumulation of excess Ca ions intracellular, which in turn initiate the calcium-driven cell necrosis.

4.2. Nozogenic Activation of Pulmonary Vasoconstriction

Its precursors are venous hypoxemia and ion-osmotic deformity of cell mass and acidosis.

Hypoxic pulmonary vasoconstriction (HPV) it is a body response to alveolar hypoxia. It manifests by a marked narrowing of the precapillary pulmonary arteries and arterioles adjoining the alveoli. In pneumonia, or an atelectasis case, the HPV has localised character, diverting blood flow from hypoxic areas and reducing the degree of ventilation-perfusion mismatch (shunt).

However, in diffuse lung diseases, the HPV affects the whole lung.

When, regardless of aetiology, vascular pulmonary reserves are shrunk, then increased cardiac output it leads to pulmonary hypertension.

The biggest trigger for HPV as we state is alveolar hypoxia. However, systemic arterial hypoxia exacerbates local effects of an alveolar one indirectly through a sympathetic pathway.

Severe metabolic acidosis (Ph < 7.2) is also able to cause HPV, acting synergically to hypoxia.

The emerging excess of water in extracellular lung areas is the result of the lack of compensatory mechanisms that are responsible for fluid entering and reabsorbed in lung tissues.

The anatomy of the alveolar-capillary membrane is shown in a Pict. 4.1

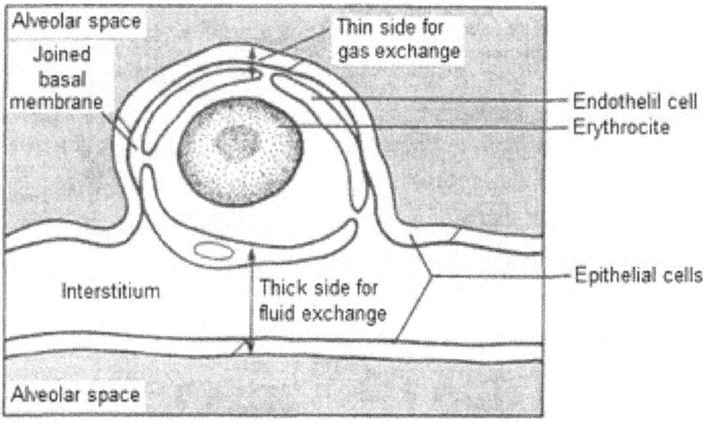

Pict. 4.1 Anatomy of alveolar-capillary membrane

Practically all alveolar interstitial tissue makes up one side of a capillary, it is the so-called 'thick side'. It contains collagen and elastin fibres, which build the structural framework of the alveolar walls. This promotes fluid exchange in the microcirculation. Opposite to it, the endothelial membrane 'merges' directly with the epithelial basal membrane, forming an extremely thin layer: the 'thin side'. This one ensures fast exchange of O_2/CO_2 through a diffusion-adapted alveolar-capillary membrane. It is likely, the most of a fluid entering the walls of the pulmonary microcirculation passes through gaps at the junction between endothelial cells (Fig 4.2.2).

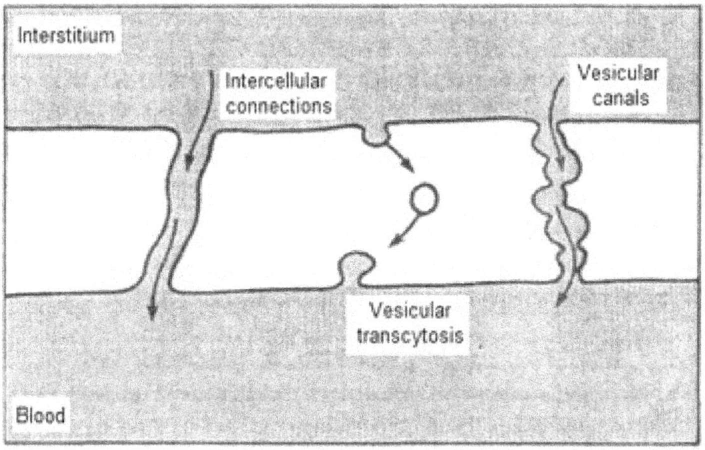

Fig. 4.2 Endothelial fluids pathways

Water, low-molecular solutions (for example glucose), and small plasma proteins (e.g., transferrin and albumin) are moved passively through by convection and diffusion.

Microvascular filtrate under the hydrostatic pressure gradient is directed from the alveolar walls to the interstitial space of pleura and broncho-vascular trees.

Terminal vessels of the pulmonary lymphatic system are located in sub-pleural and peri- bronchial connective tissue.

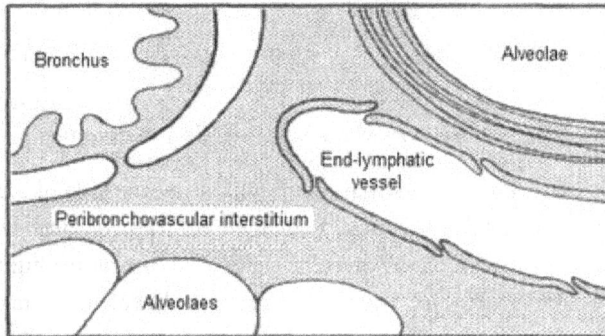

Fig 4.3 Terminal lymphatic tree

The fluid enters the lymphatic junctions through a gap in the lymphatic endothelium. It pumps with peristaltic waves, bypassing a series of one-way valves to collective lymphatic vessels by the lung's roots. From there the fluid goes through the mediastinal lymph vessels to associated lymph nodes to the common thoracic duct, and then returns to a blood. This pathway is schematically shown on Figure 4.2.4

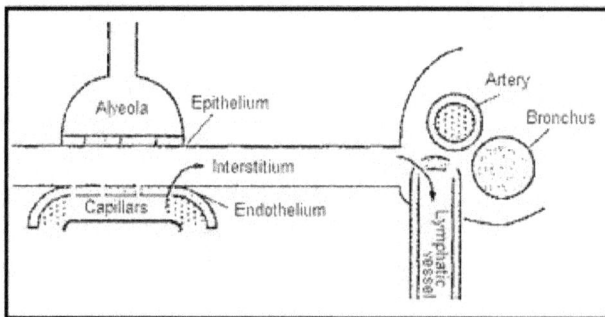

Fig 4.4 Scheme of fluid exchange in the lungs

Pulmonary oedema occurs when fluid is filtered through the pulmonary vessels faster than it is reabsorbed by the lymphatic system. First, fluid accumulates in the interstitial space (interstitial oedema), causing slight 'thick side' swelling, filling up pleural interstitium and spaces around bronchial bundles. The latter are more stretchable than similar in the alveolar walls. In fact, the subpleural and peri -bronchial connective tissues serve as a 'runoff' that carries excess water from the alveoli, protecting the lung gas exchange. The plural spaces act as a second 'reservoir' which removes fluid from lung interstitium.

If a large amount of fluid enters the pulmonary interstitium quickly, or alveolar epithelium is damaged, the fluid penetrates the air space and fills alveoli (alveolar oedema). When alveoli are filled, the foamy fluid enters the bronchi. At this stage, the gas exchange is significantly impaired, a big shunt develops as blood bypasses damaged alveoli now. Lung elasticity and compliance is markedly reduced.

4.3. Hypoxic Ischaemia Reperfusion Lung Injury

In case of ischaemia not only oxygen delivery suffers but the other metabolic component transport (e.g., glucose, fatty acids, proteins, electrolytes) is disrupted. It combines hypoxic-metabolic insufficiency. The most sensitive to these are CNS neurons, cardiomyocytes, transverse stripped muscles, smooth vascular muscles.

In hypoxic ischaemia, aerobic respiration and oxidative phosphorylation in mitochondria is rapidly impaired, causing decreased ATP production. In a short period of time compensatory anaerobic glycolysis is activated. As a result, there is increased concentration of lactic acid and hydrogen protons, leading to metabolic acidosis developing.

In acidosis, glycolysis is exhausted and total energy deficiency of cells rapidly increases.

- A severe consequence of energy shortage is a decrease in membrane potential of CNS neurons and cardiomyocytes (these are in normal conditions spend 25–30% of its energy for a potential maintenance).

- The system of energy-dependent transport of ions and a water through the plasma membrane is disrupted and ischaemic swelling of cells occur as interstitial water enters a cell.
- During progression of hypoxic ischaemia, the enzymatic lipolysis and proteolysis are activated causing decomposing of cell membranes and cell organelles lysis. In addition, as the energy biosynthetic process is suppressed, the intracellular repair is grossly compromised.
- Through the damaged membrane, extracellular calcium moves and accumulates inside of a cell, triggering a calcium-dependent mechanism of its destruction.

Reperfusion damage occurs when the blood circulation resumes after prolonged ischaemia of an organ. Reperfusion has following features:

- It develops not only in the cells of the ischemic organ, but also in organs that did not suffer complete ischaemia. It happens through the remote actions of cytokines and toxins of damaged cells transmitted by blood.
- It launches a cascade of delayed cytolytic, thrombo-haemorrhagic and hemodynamic disorders that will last a few days after ischaemia

At the molecular level, the reperfusion alteration manifests with:

- The free O2 radicals as well as the peroxidation of lipids, proteins and carbohydrates formation.
- Calcium cascade of cell necrosis and apoptosis.
- Endonuclease fragmentation of nuclear DNA chromatin and following cell necrosis.

Despite all the above, each cell has evolutionary formed mechanisms, which are timely to eliminate or minimise molecular metabolic damage.

Modern molecular research suggests that the neurohumoral response on reperfusion and systemic damage of oxygen-resource supply is a component of unified reaction of neuroimmune system, caused by expression of genes of early activation, synthesis of stress modulators and cytokines.

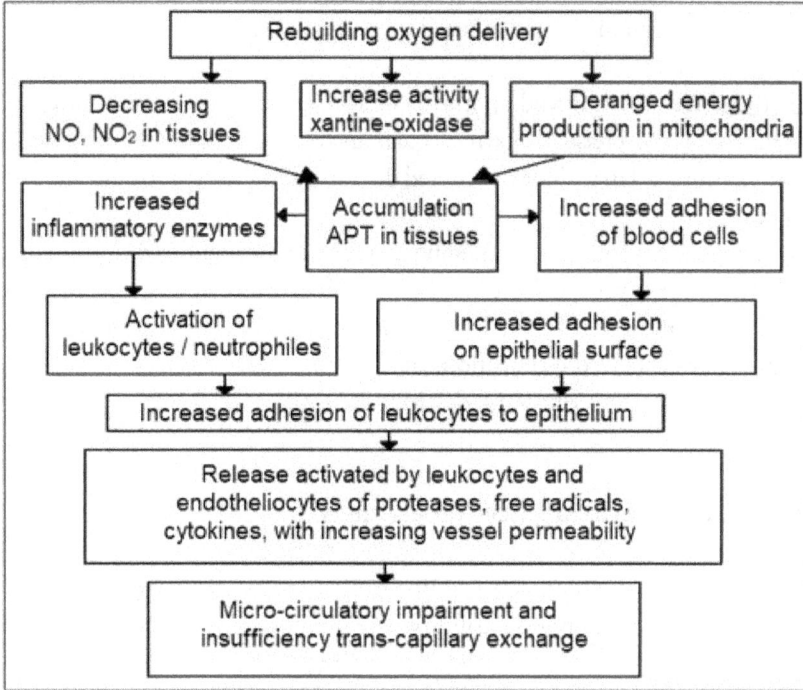

Fig 4.4 Pathogenesis of reperfusion

Releasing activated by leukocytes endothelial proteas, O2 free radicals, cytokines, which increase vessels permeability.

Microcirculation impairment and dysfunction of transcapillary exchange.

Reaction on pathogenic influence ensures energy providing a basis for compensatory and adaptive changes, carried out by sympatho-adrenal and thyroid gland systems.

Pathogenically induced apoptosis, unlike one in physiological conditions, is a type of selective death of specialised cells, which is prematurely initiated by genes in case of cell molecular damage, while this cell is not yet ready to die. Under pathological conditions, external and internal molecular damage can cause apoptosis in any phase of cell cycle. They either react with transmembrane receptors or activate molecular signalling pathways in cytoplasm to cell genes, initiating cellular death.

In the case of critical damage, apoptosis is a gene-regulated method of rapid self-liquidation of damaged cells by cascade activation of caspases and endonucleases with disintegration of its residues by macrophages. Also noted is white cell death, which is premature death of a cell that exhausted its survival programs, implemented by calpains, lysosomal acidic proteas and hydrolases.

Pathogenetically induced apoptosis of specialised cells is morphologically diagnosed in the phase of its molecular initiation, in the second phase of inflammation and in a phase of ordered disintegration (apoptotic degradation) of a cell.

Intermediate products of metabolism of the oxidant stress system, as secondary messengers, initiate a series of life-saving reactions in ischaemic tissues. This phosphorylation serves as a signal to the early gene's response, which even occurs in the conditions of suppressed (due to violation of amino acid transduction and ribosome destruction) protein synthesis.

4.4 Capillary Permeability Impairment

One of the reasons for pathological influence, due to insufficient oxygen supply, is a pathology of capillary permeability. This mainly depends on the functional state of endothelium and microcirculation. The nosogenous damage disrupts the balance of vascular tone regulators, such as catecholamines, angiotensin, etc. In addition, a damaged endothelium loses anti-inflammatory, antioxidant, antithrombogenic properties. This contributes to the activation of parietal thrombosis, perfusion disorders and tissue hypoxia.

Changes in the functional state of erythrocytes play an important role in restricting capillary blood flow, cause decrease in elasticity of the surface membrane and increase the tendency of erythrocytes to aggregate. It should be recalled that the main functions of blood cells are performed in capillaries.

In the capillary network, the diameter of erythrocytes corresponds with the diameter of a capillary, and only due to elasticity of the surface membrane of erythrocytes it is possible to move blood cells in capillaries. At pathological destruction, the lipids composition of erythrocyte membrane changes, increasing a membrane stiffness, which significantly complicates and sometimes blocks the passage of an erythrocyte through the capillaries. In addition, the change in a membrane structure leads to a fall in the negative charge of an erythrocyte surface. This causes adsorption of colloids of plasma on a surface and its subsequent denaturation. The resulting film, which consists either of fibrinogen or gamma or beta-globulins, causes subsequent adhesions of thrombocytes.

Aggregated erythrocytes aggravate the tissue hypoxia and also trigger the activation of intravascular thrombosis by erythrocytes capillary blockage.

The role of hemorheological disorders in the pathogenesis of microcirculatory ischaemia is not limited by mechanical properties of damaged erythrocytes. It is known that if the vascular diameter twice as less there will be 8-th fold increase in a blood flow rate, and thus leads to increase in the shear stress on vascular endothelium. The shear stress in aorta and large vessels fluctuate from 2 to 11 din/cm2, while at the arterial (microvascular level) it is 60 din/cm2. It is clear that microvascular endothelium and the bloodstream elements are under significant pressure, which even with a little increase in blood viscosity can lead to damage of blood cells and activation of intravascular thrombosis.

In the microcirculatory disorders as well as in reperfusion injuries the role of neutrophilic granulocytes cannot be overlooked. The latter size corresponds with a vessel diameter and adhesion of leukocytes to the surface of venular endothelium also can lead to capillary blockage. The more neutrophilic granulocytes are involved in the reaction, the more significant the disturbance of capillary blood flow, the greater the consequences.

In addition, neutrophilic granulocytes are the source of biologically active substances, cytokines, free radicals, the release of which significantly increases the permeability of the endothelium. This promotes the development of local oedema and activation of thrombosis. Migration of neutrophilic granulocytes and monocytes into the damaged endothelium occurs within the first 15 minutes after an induction of cytokines and histamine, and within two hours there is a leukocytes storm, its adhesions in venules and its trans endothelial deposition.

Changing the functional state of platelets plays a significant role in the disturbance of intravascular homeostasis, as thrombocytes are the central link of blood coagulation. And although the mechanical properties of platelets, due to their small size and limited number, play a minor role in aggregation process, the thrombocytes sensitivity to procoagulants may cause significant impact on a state of blood flow at the level of microcirculation.

As a result of destabilisation of intravascular homeostasis, a violation of protective properties of microvascular endothelium, and an increase a blood viscosity in the conditions of activation of substances causing adhesion of blood form and aggregates formation, there is a significant shutdown of blood flow in micro vessels (blood stasis). And the final development of a sludge-phenomenon characterised by adhesions, aggregation and agglutination of blood elements is the development of capillary trophic failure syndrome.

4.5 Energy and Cellular Energy Resuscitation Impaired Energy Structural Coupling

The aim of cellular energy resuscitation is the restore a vitalism, as without this restoration of the mechanisms of self-reparation of BCM, survival of any organism is impossible.

This technology should minimise energy losses, eliminate the possibility of oxygen debt development, as well as intensify the production free energy flow, necessary to restore BP in a situation of nosogenous damage of energy structural conjugation. The indicator of rectification of matching intensity of a process to the organism's needs in physiological regeneration can be SvO1 of mixed blood. This, in iso-Osmia practically corresponds with the average tissue oxygen saturation.

The leading role in reaching the vitalism reserve is played by anti-energy consuming energy-osmolar optimisation. Its essence is to bring the current osmolality of plasma to a level that minimises energy expenses. Then the energy cellular osmolality deficit/excess is determined, and the total amount of osmolality required for optimisation is set.

Stabilisation of energy-structural disbalance at loss of vitalism is accelerated if the values of the quantum biocycle (QBC) is transformed into the achievement of HR, determined by the equation of Kerdo:

$$\text{Stabilising QBC HR} = \text{DAP}/0.93, \text{ mm HG}$$

where:
DAP – diastolic BP, HR – heart rate

Oxygen-hemodynamic harmonisation has a vitalotropic effect on the energy-structural status as it able to exclude hypoxic or hypertoxic damage of BCM. It is achieved by the ability to synchronise the rate of erythrocyte deoxygenation with the level of normotension (from AD: 105/60 to AD 140/90 mmHg).

Conjugated energy-structural interaction will become inherited to the vasomotion if the rhythmicity of its function is controlled by the level of HCO3-, which corresponds with the intensity of tissue metabolism. Its relevant value can be obtained from the formula used to calculate an anionic gap:

$$HCO3\text{- gap} = (Na+ Cl) -12, \text{mmol/l}$$

where:

Na – sodium concentration in plasma, mmol/l

K – potassium concentration

12 – average anionic gap, mmol/l

Energy cellular vitalism resuscitation requires reliable energy protection; therefore, it must be preceded.

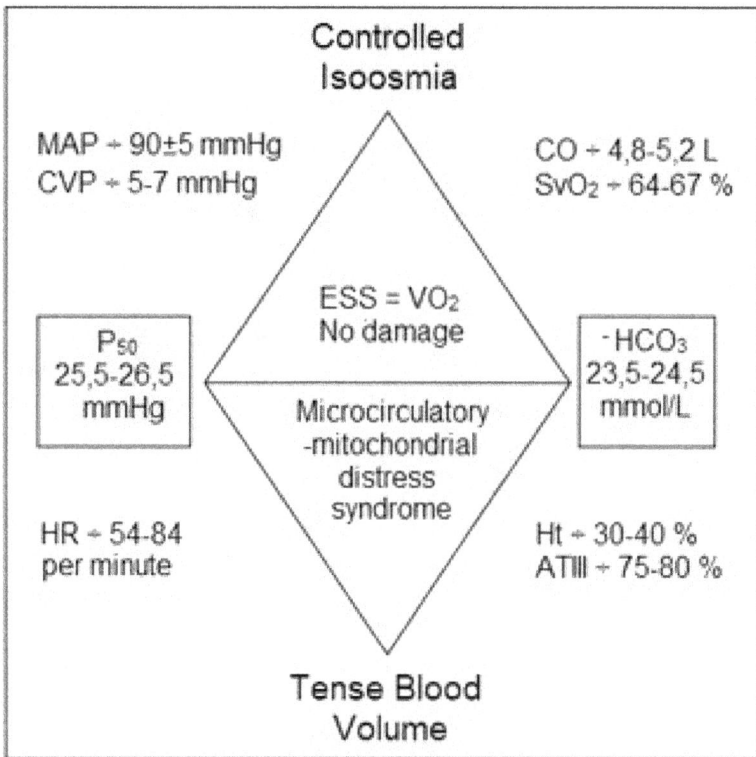

Figure 4.5 Energy resuscitation scheme

The key to an efficiency of organism's gas exchange system is Guyton's 'stressed' blood volume Vs, which is the basis of blood flow. It determines a venous return to a heart, the degree of stretching of the microcirculatory channels, required to maintain vasomotion. Therefore, Vs ensures economical heart work, corresponding with the Frank–Starling Law, and also responsible for seeing that microcirculation intensity is optimal. Fluid boluses with CVP monitoring have a great diagnostic value, especially in hypotensive patients and the patients with impaired cardiac function. If the heart inotropic function is normal, the bolusing of fluids results in transit ABP increase without raise in CVP, which suggests the patient is hypovolemic. However, in cardiac failure, CVP rises with no recordable SAP increase. Therefore, an absolute contraindication of fluid boluses is an increase in pulmonary pressure above 18 mmHg.

After reaching Vs, the infusion therapy is calculated from the equation:

$$\textbf{(20ml + insensible losses + diuresis) x 10 min}$$

The stress volume, providing optimal preload on a heart allows the heterometric mechanism of the heart to be run to the maximum possible extent without maximum energy being used.

The measures for achieving stress volume of a heart listed below:

BP	CVP	Measures
N	N	Usual intensive therapy
N	↑	Intensive therapy + diuretics to keep TBV
↑	N	Use B-blockers with a control TBV
↑	↑	Intensive therapy with drugs reducing venous return
↓	↑	Cardiogenic shock: rule "5-2"
↓	N↓	Fluid boluses to maintain TBV

Fig 4.6 Stress blood volume measures

Washed red cells, leukocytes free red cells, used for Hb restoration and maintenance are a good oxygen transport.

All blood products need to be transfused through special filters which trap microunits blood cells, leukocytes and fibrin

… The Hb equal or below 70 g/l should be considered ultimate.

The required red blood cell (RBC) volume is determined by a formula:

$$\textbf{RBC = 5 x (70 – Hb) x CBV, ml}$$

where CBV: circulating blood volume (CBV = 0.08 x BM (kg))
For a sample: BM = 50 kg, Hb 50g/l
RBC = 5 x (70-50) x 4 = 400 ml

Getting a diuresis of 0.5 ml/kg/h or more will indicate that the bioenergy deficit is tackling and oxygen debt eliminated. The clinical symptoms and functional signs of improvement would be effective acidosis clearance and active removal of H+ from a urine (acidotic urine PH: + and <6.5).

This will be followed by an energy cellular resuscitation to restore the reserve of the vitalism.

4.6.1 Energy resuscitation scheme

The hypoergia is presented, respectively, by the extensive shutdown of damaged body mass cell, which reliably reflects the deficit of vitalism, i.e., 'vitability' (Table 4.2).

As can be seen from Table 4.2, the energy and energy – cellular resuscitation begins at a stage of hypobiotic insufficiency, which possess significant dangers. The expansion of the damaged body cell mass will be followed by the exhaustion of oxygen transport and microcirculatory and mitochondrial reserves.

Characteristics of Vitalism	Base-line of hypo-energy%	Energy resuscitation, %	Energy-cellular resuscitation, %
Reserve of Vitalism	÷	÷	2,6
Deficit of Vitalism	19,3 ÷ 1,01	18,6	÷
Myocardial reserve	46,6 ÷ 54,4	52,2	57,7
Oxygen transport impairment	26 ÷ 6	26	11
Micro-circulatory-mitochondrial insufficiency	21,4 ÷ 4,4	20	9,8
Osmolar stability	÷ 0,3	0,9	0,7
Hyper-osmolar destability	1,9 ÷	÷	÷
Haemodynamic stability	÷ 1,48	÷	÷
Diastolic destability	18,1 ÷	10	2,1
Adaptivity	÷ 118,1	103	121
Destructivity	0,4 ÷	÷	÷
Unstability	21,7 ÷ 5,3	21,1	9,7
Adequacy	÷ 112,8	÷	111,3
Inadequacy	22,1 ÷	10,8	÷

Table 4.7 Properties of energy and energy-cellular resuscitation during hypoergic instability

Please note, energy-osmotic autoregulation is preserved, which stabilises the cytoarchitectonics. Diastolic dysfunction is the result of poorly controlled autoregulation of hemodynamic protection. If even the destruction of body cell mass is prevented, the energy-protective activity remains inadequate by 10,8% (p < 0.05).

Therefore oxygen-hemodynamic therapy included a maintenance of BP within the range of 105/65 to 120/75 mm Hg, and the HR 65-80/min respectively.

After energy-structural delivery is matched with the energy consumption of body cell mass, the energy-cellular resuscitation begins, leading to a notmobioty, although some deficit of vitalism is in evidence. This can be seen in moderate depletion in oxygen transport and microcirculatory and mitochondrial reserve.

Therefore, in the treatment scheme energy substances were used, with a potential of 805–1000 kkal/24h.

4.6.2 Energy and Energy-Cellular Resuscitation in the Catabolic Impairment

Catabolic damage determines the production of active oxygen forms (AOF), the intensity of which can be compared to overflow, when hyperergia significantly exceeds the upper level of energy production reliability. (Table 4.3)

Characteristics of Vitalism	Base-line of hypo-energy%	Energy resuscitation, %	Energy-cellular resuscitation, %
Deficit of Vitalism	30,6 ÷ 55,4	41	19
Myocardial reserve	51,1 ÷ 38,3	30	47,7
Oxygen transport impairment	15 ÷ 30	20	10
Micro-circulatory-mitochondrial reserve	15,2 ÷ 26,9	20,2	9,7
Hyper-osmolar destability	1,8 ÷ 8,5	4,6	1,4
Systolic destability	22 ÷ 11,7	6,2	4,7
Adaptivity	142,3 ÷157	147,8	136,8
Stability	115,1 ÷127	120,1	109,5
Adequacy	158,4 ÷184	167,9	146,3

Table 4.8 Properties of energy and energy-cellular resuscitation during hyperergic instability

As seen from Table 4.3 the hazardous consequences impact reached with the catabolic damage of 1/2 -1/3 of amount of BMC.

In this case the oxygen consumption for ATF reaches 37–103 ml/min x square meter. However, with haemodynamic stabilisation (BP 150/95 – 160/105, HR 95-110/min), cardiac, microcirculatory and mitochondrial reserves are preserved, increasing with it the efficiency of energy resources, adding timely calcium channel blockers and maintaining a normothermia.

The difficult task of osmolality correction should not be solved by total elimination of sodium deficit at hypo-osmolar destabilised body cell mass. The energy-osmolar optimisation should be at the range of osmolality of 290–294 mOsm/kgH2O. and oxygen consumption 174 – 190 ml/min x square metre.

During controlled hyperergia, it is appropriate to maintain good urine output, and avoid a manifestation of hypovolemia as this affects cytoarchitectonics.

As the result, the catabolic instability of body cell mass (BCM) is illuminating reliably, and the possibility of dangerous energy-structural disorders with residual vitalism deficiency vanishes.

Preservation of energy-structural reserve emphasises the energy protective capacity of energy-cellular resuscitation.

4.7 Reparation Of Damaged Cells

By restoring the vitalism reserve DNA repair is facilitated. The coded in specific nucleotide sequence the genetic information is in each of two DNA helix chains. The damaged nucleotide in one of the chains can be removed by the reparation enzyme, and as the information is stored in an intact DNA chain, the new nucleotide is created as a copy of the impaired one. Full regeneration of DNA, such as matrix repair, is completed in 24–30 hours from the impact of a damage.

Mitochondria alone are self-renewable organelles with a solitude genome and brief life cycle: half-life of cardiomyocytes mitochondria is about 6.5 days, hepatocytes ones: 9.6–10.2 days, renal cells: 10–12.5 days, and neuron mitochondria lives around a month only. The DNA of mitochondria is a structure of a double-chain covalently closed ring, attached to the inner membrane of the organelle. It is distinguished by superior spiralisation, absence of introns (non-coding region of RNA transcript), and a limited amount of stored genome. In addition, it encodes the synthesis of three mitochondrial RNAs (messenger, ribosomal and transfer), the synthesis of 13 of 70 existing polypeptides of transport chain of electrons, as well as 2% of own proteins and many other enzymes (except of participants of respiratory ones, together with some subunits of ATP-synthesis cycle. (Gayze)) Autonomy of mitochondria in the cell is limited: nuclear DNA controls replication of mitochondrial DNA and code a synthesis over 90% of ribosomal proteins, enzymes of citric and pyruvate acid cycle, transfer and oxidation of fatty acids and majority of polypeptides of transport electrons.

In contrast to nuclear DNA, the synthesis of mitochondrial DNA is intense in multiplying and interphase cells, which is associated with constant renewal of organelles.

It is established that each mitochondria contains 2–10 copies of mitochondrial DNA, and a single cell accommodates up to 10,000 copies of its kind.

Essential for glycocalyx and membrane proteins as well as enzymes for organelles are synthesised by ribosomes attached to the outer membrane of endothelium.

New synthesised proteins undergo differentiation, after this the proteins either remain in the cytoplasm, or transported to nucleus mitochondria, and endoplasmic chains.

The info of each protein (its code) is located in its amino acid unique sequence. If newly occurring proteins did not go through maturation they remain in cytoplasm of a cell, where they will perform various enzymatic functions or will be modified to take a part in plasmalemma updating.

The proteins which entered the endoplasmic reticulum after being modified then 'matured' in the Golgi complex, ready for next sorting. Some proteins then enter lysosomes and peroxisomes, others go to secretory granules and glands, or become transfer ones.

Changeover of membranes, lipids and proteins, as well as glycocalyx is possible with the help of transfer proteins and transfer bubbles, later full of coded pores to be able recognise and accept particular enzymes in specific places of membrane. Transfer bubbles contain proteins specifically modified in the endoplasmic net for a membrane of other organelles. During the transfer the bubbles are detached from the Golgi complex and merged with the matching compartment of the particular organelles. The membrane phospholipids, in its turn, are synthesised on the cytoplasmic side of the endoplasmic reticulum and then get transported to the membrane by transfer proteins

Proteins intended for the regeneration of nuclear structural components are advanced to karyoplasm through the nuclear pores, which speed up the membrane transport. Specific organelles of a cell's catabolic reactions (lysosomes, peroxisomes, exocytosis-pinocytosis vesicles) are reproduced according to the metabolic needs.

Lysosomes are synthesised from endoplasmic reticulum through the intermediate product – endolysosomes. The later accept transported proteins bubbles from the Golgi complex while hydrolyses come from cytosol, in clathrin-covered bubbles. The transfer initially forms lysosomes, holding the necessary set of hydrolases (aka phagosome) to where the hydrolysis is taking place, which is carried out with a help of microtubules of cytoskeleton. After the fusion of the primary lysosome and phagosome, hydrolases are activated, causing a melting of the phagolysosome.

Products of lysis enter cytoplasm for further metabolism, and phagolysosome becomes a secondary lysosome that comes to contact with the next phagosome again.

After a few such cycles, residual hydrolysis products accumulate inside the secondary lysosomes. The latter, then losing its hydrolytic abilities, are transformed into residual bodies which are either preserved in the cell or removed by exocytosis.

Peroxisomes, similar to lysosomes, are formed from the endoplasmic reticulum as needed. They become functional after a selective intake of specific enzymes, which come from cytosol. Peroxisomes hold around 40 subtracts, which play a significant part in inactivation of free radicals formed in a cytoplasm (such as atomic O_2, peroxidation and hydrogen peroxide). These 40 enzymes are also responsible for metabolism of fatty acids, cholesterol, bile acids, purines and oxalates, in the gluconeogenesis processes, as well as degradation of prostaglandins, some medications and xenobiotics

CHAPTER 5

Status-Resuscitation in Vitalism Failure

5.1. Evolution of Nosogenous Implementation of Energy-Structural Interaction

Evolution is the result of changes in certain genes' expression. For instance, the mediator-cytokine reaction can be represented as a consequence of an imbalance of genetic regulation of the mediator's response, which can make qualitative changes in the metabolic processes during noso-realisation. Mechanisms of induction and repression prevent the cell from wasted amino acids and energy, and it's all determined in the cell's genome.

The induced control of genes of cytokines (at least TNF), cytokine receptors, the main complex of histocompatibility antigens, acute phase proteins, and – according to some data – macrophages nitrogen oxide, is done by the nuclear protein kB/NF-kB (synthesises in nuclear matrix as a factor of transcription). Moreover, the induction of m-RNK cytokines (TNF, IL-1) requires activation (phosphorylation) of the protein kinase C and protein tyrosine kinase. It leads to proteolytic lysis of the inhibitor form and subsequent initiation of transcription of the corresponding genes. Nk-Kb can be represented as an inhibitor in the presence of protein tyrosine kinase as an inductor.

This example highlights the importance of energy-structural coupling, responsible for combining the genes of various metabolic pathways.

Recognition of the continuity of the development of pathological processes allows for early recognition of a danger of vitalism exhaustion and allows one to carry out necessary treatment measures. The biggest prognostic factor in this is the excess production of cytokines and other mediators of inflammation, which can disrupt immunological regulation of preservation of vitalism.

Pro- and anti-inflammatory mediators can eventually reinforce each other, creating a growing immunological dissonance. There is a direct correlation between the load of cytokines in the blood, the severity of splanchnic ischemia and hypoxic damage to intestinal mucosa. Since the intestine is the breeding ground for a bacteria and toxins, it can become a powerful motor of energy-structural disorder.

Intestinal damage (ischaemia/reperfusion) causes delayed organ failure through the activation of circulating neutrophils. They become activated in intestinal blood flow by the end of the second hour of reperfusion. This stage precedes neutrophils systemic activation, which makes an intestine the source of inflammation. It is especially important that all processes occur independently of the endotoxin impact.

The biological phenomenon when the primary damage only prepares the 'host' body for an overreaction to secondary damage is called by Edwin Deith (1992) as the 'phenomenon of two strikes.' Shock leads to tissue ischaemia, against which the 'host' body modulates the response to subsequent damage (bacteria/endotoxins). For example, in polytrauma, during a hypotension, a decrease in blood flow in organs may lead to a clinically unidentified area of ischaemia/reperfusion damage, initiating inflammation.

Subsequent damage done by bacterial/toxins translocation will result in increasing tissue response/damage.

Oblique anaerobes, responsible for colonic bacterial antagonism and colonisation resistance of the body, form the first line of defence or barrier of intestinal mucosa from pathogenic microorganisms. Violation of the intestinal ecosystem and inadequate oxygen supply leads to decrease in both numbers of obligate anaerobes and their individual representatives in small and large colon, especially lacto-bacteria. The latter are known to hold antimicrobial properties.

Bacterial translocations undergo two pathways:

- Mesenteric lymph nodes ----- thoracic lymph duct -------- systemic blood flow (6–12 h)
- Portal vein ----------------------- liver ---------------------------systemic blood flow (12–24 h)

The above mechanisms are implemented through the activation of macrophages, their secretion of mediators and phagocytosis disorder.

Nosogenous implementation mechanisms that are responsible for energy-structural disintegration, are the main reason for various disturbances and damages. The main scheme of organ failure in the total body and deficit of vitalism is presented in Fig 5.1

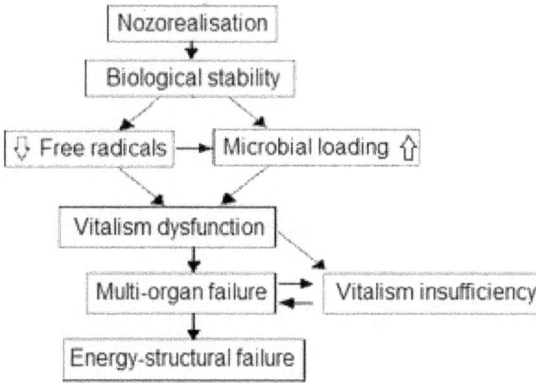

Fig 5.1 Development energy-structural insufficiency in nosogenous damage of conjugation

The endothelium exposed to cytokines enhances:

• The surface adhesion of E-selectins molecules
• Intercellular adhesions of ICAM -1 molecules
• Adhesions of VCAM (vascular cell adhesion molecules) vessels to attract polymorphonuclear lymphocytes and monocytes to adhere and penetrate the source of infection
• Change functions of activated endothelium to procoagulant and antifibrinolytic module

The most frequent external factors are ischemia, hypoglycaemia, reperfusion, gamma and ultraviolet rays, molecules of intracellular interaction (TNF-tissue, lack of growth factors, excess glutamate), as well as hormones, and antiviral drugs, viruses, molecules released by macrophages, sensebilised lymphocytes in the contact with the target cell.

At death by apoptosis, the cell will disappear in 1–3 hours.

During the first 10–60 minutes the molecular changes occur, which could be identified by molecular immunocytochemical methods.

Apoptosis in critical injury is generally controlled way of rapid self-liquidation of the initially damaged cells, while necrosis is the premature death of cells that have exhausted all their survival programs.

In the phenomenon of cell necrosis, damage to the plasma or mitochondrial membranes, endoplasmic reticulum and lysosomes play a crucial role. The necrosis is accompanied by advanced swelling of the cell and its organelles, karyo-lysis or karyo-pyknosis, lysis or condensation of cytoplasm, near-cell accumulation of macrophages and neutrophils. It also does not require the synthesis of new proteins and uses up minimal cell energy. Both necrosis and apoptosis in disease may affect the identical cells (CNS neurons, hepatocytes).

Selective cell death induced by the immune system and involved two processes: recognition by the immune cells as 'alien' and their immediate destruction.

In this way, infected cells, tumours, transplanted cells, or normal but recognised 'alien') cells (in autoimmune diseases), are destroyed.

Immune destruction of the target cells is done either by immunocytes or by activated complement and occurs through phagocytosis or immune-cellular killing.

5.2 Main Pathways of Status Correction

Eliminating the failure of vitalism should be, first, aimed at the correction of the transport and delivery components of the oxygen regime, which are the weakest link of BMC at this stage of damage. Therefore, the highest goal of status-correction should be the elimination of reversible pathogenicity, responsible for torpid insufficiency of biostability. Its distinctive feature is torpid induced arterial hypotension when the minimal perfusion pressure is difficult to achieve. This is a pathognomonic symptom of shock syndrome: arterial hypotension in combination with signs of tissue hypoperfusion. The definition of arterial hypotension in normotonic state: systolic BP below 90mm Hg or decrease BP by 40 mmHg in hypertonic state.

Tissue perfusion and oxygenation disorders in shock, combined with microcirculation disorders, make it impossible to meet the oxygen demand of tissues. Haemorrhagic shock results from bleeding (hidden or visualised) First of all, it is necessary to detect and eliminate the bleeding, then to start iso-osmolar oxygen-protective infusion therapy (usually both done at the same time). Compliance with the rule is a fundamental moment in the treatment of haemorrhagic shock, and crucial in liquidation of a vitalism deficiency.

Haemorrhagic shock is diagnosed in various pathogenic conditions, such as an absolute decrease in the volume of body fluids, or the redistribution (shunting) of extracellular fluid from intravascular compartment to the interstitial (third) space.

Heart tamponade is considered the most common cause of obstructive shock. The leading role here is played by the rate of fluid accumulation in the heart sack (pericardium), not its quantity. The main causes of cardiac tamponade: myocardial infarction and aortic dissection. Clinical signs of a tamponade: high CVP, protruding jugular veins, strong BP fluctuation in breathing. Obstructive shock is also a feature in a tension pneumothorax. Later and a tamponade can be seen in ECHO screen, its main diagnostic method.

The main mechanism of cardiogenic shock formation is the primary reduction of the cardiac pumping function, caused by myocardial infarction, cardiomyopathy, critical aortic stenosis, severe aortic effect of mitral valve insufficiency, cardiac arrhythmias or large inter -septal defects.

Often cardiogenic shock is caused by the heart rhythm disorders. Clinical studies showed that severe AF/VF and bradycardia/heart blocks result in systemic hypoperfusion and cardiogenic shock, with equal frequency.

In order to obtain the classic hemodynamic profile in septic shock (high CO, low CI, together with systemic arterial hypotension), it is necessary to look for a source of infection and start antibacterial therapy early in every patient.

Another type of shock is spinal shock. It is a manifestation of distributive shock, which occurs in traumatic spinal cord injury above the mid-thoracic region. Clinical symptoms are arterial hypotension, bradycardia, dry and warm skin. However, if you suspect spinal shock, you must exclude other causes of haemodynamic instability (like latent bleeding). In spinal shock there is found a relative mismatch between the capacity of the vascular channel and the volume of circulating fluid. Therefore, the first step is an intravenous infusion of iso-osmolar crystalloids. If a hypotension does not respond to a fluid, inotropic support should be implemented. However, the use of the volume can overload the volumetric load of the heart, and the use of vasopressors will inevitably increase the afterload of the left ventricle, this will exacerbate heart failure. In addition, the use of inotropes constantly will also increase the heart oxygen requirement, and with insufficient oxygen delivery (due to low coronary perfusion pressure) if the demand is not met, heart failure will progress.

The cardiac fragment of status correction is presented in Table 5.1

Causes	Treatment
Inconsistency of energy consumption	Provision of Tense Blood Volume (TBV) under the rule "5-2"
Volume overload, peripheral distress syndrome	Low volume, normocarbic ventilation
Myocardial insufficiency	Maintain TBV, inotropic support, treatment hypoxic vasoconstriction
Tachysystole	Treat coronary hypoperfusion (Diastolic BP 70-80 mmHg), potasium normalising, B-blockers

Fig 5.2 Correction of coronary perfusion

The activation of anaerobic glycolysis may be seen as a compensatory reaction, directed to dealing with increasing energy demand in hypoxia. Glycolytic energy production greatly helps support the Ca- pump of reticulum, the transfer of macromolecules to contracting proteins by activation of cytoplasmic phosphokinase, the preparation of amino acids for the tri-carbonic acids cycle, and the maintenance of the myocardiocytes action potential. ATP deficiency significantly increases a risk of development ventricular fibrillation. Modelled in experimental conditions it has been proven infusion of glucose with maintains of glycolytic ATP production normalises the stability of an action potential (Horn, 1999). Also, glycolytic ATP production associates with the synthesis creatine kinase in cytoplasm.

Endocardial survival rate (ESR) plays a crucial role, it has to be ranged from 1.2–1.4. And inotropes also help provide such ESR values.

5.3 Energo-Protection in Acute Lung Injury/Acute Respiratory Distress Syndrome (ARDS)

Acute lung injury syndrome is pathognomonic in biological body destruction. The development of ARDS mostly occurs in the first 12–48 hours of the onset of the pathogenic factor's impact. Growing vitalism deficiency is based on damage to epithelial and endothelial lung barriers, acute inflammatory response and the development of pulmonary oedema. All the above leads to acute respiratory failure.

The genesis of pulmonary hypertension in ALI/ARDS is highly multifactorial: hypoxic vasoconstriction, vasospasm caused by vasoactive mediators of inflammation such as thromboxane, leukotrienes and endothelin, as well as intravascular obstruction with thrombocytic clots, and inevitable perivascular oedema

The main feature of ARDS is hypoxemia (SATS below 90%–75%). ARDS patients are almost refractory to oxygen therapy, which reflects the main mechanism of gas exchange disorder due to the development of intrapulmonary shunt. In the early stages of ARDS, hypercapnia ($CO_2 > 45mmHg$) and respiratory acidosis ($pH < 7.35$) are associated with high minute ventilation. As the condition progresses, there is an increase in alveolar dead space, poor CO_2 clearance, respiratory muscle fatigue and the development of respiratory alkalosis (replacing an earlier acidosis)

The analysis of bronchoalveolar lavage (BAL) contents in the first days of ARDS reveals the high number of neutrophils (over 60%), which if a treatment is successful will be replaced by macrophages.

A distinctive radiological feature of ARDS is the 'ground glass' appearance of the chest x-ray, and diffuse multifocal infiltrates of rather high density (consolidation), with well-defined airborne bronchograms (i.e., the development of a widespread lesion of the lung parenchyma).

Although the damage is diffuse, the lung tissue is not affected uniformly: there is uneven distribution. This has significant therapeutic consequences.

There are two main etiological reasons for ARDS: primary lung disease (pneumonia) or its extrapulmonary cause, like sepsis. In the first case the damage to pulmonary epithelium is limited by single organ failure. In the second instance capillary endothelium damage occurs due to systemic inflammatory response, and lung damage becomes rather secondary and makes it one of many components of multi-organ failure. Despite the difference in pathophysiology the outcome of ARDS of pulmonary or extrapulmonary origin does not differ much.

Like every injury, an acute inflammatory response in the lungs undergoes two successive phases: exudative and proliferative. The first phase is characterised by acute damage to the alveolar epithelium, capillary circulatory failure. alveolar inflammation and oedema. Activated neutrophils release proteases, oxidants and leukotriene, while alveolar macrophages release cytokines, TNF and interleukins. The alveoli become filled with protein-like exudate, which then activates the surfactant, while the basal membrane becomes hyaline saturated. All of these ends with a diffuse alveolar collapse, intrapulmonary shunting and a decrease in the ventilatory-perfusion ratio (V/Q mismatch), progressive lungs stiffness and hypoxemia. Increasing or maintaining functional residual lung capacity is reached by implementing positive end-expiratory pressure (PEEP), which is useful in reversing hypoxemia.

The proliferative phase is characterised by reparation, resorption and scarring. The alveolar fluid of the oedema gets absorbed, macrophages phagocyte intra-alveolar proteins and neutrophils (that went to apoptosis), and pneumocytes of the second phase go through metaplasia with fibroblasts. This may result in healing or fibrosis. Alveolar integrity will be restored, while the capillary network is progressively destroyed. Inevitably, it will increase the dead space, will have high V/Q mismatch and progressive hypercarbia.

Viewing ARDS as a dynamic process with certain stages makes it possible to diagnose and differentiate the choice of respiratory support at each stage of disease in time and before severe hypoxemia develops.

The first principle of traditional ALI/ARDS is the diagnosis and a treatment of underlying causes. It is necessary to stop the primary damaging factor and prevent further stimulation of the inflammatory response of a body. Since infection and sepsis are the most common causes of ARDS, antibiotics are the first line of a treatment. In some situations, such as an abdominal sepsis or localised abscess, there is a need for a surgery to eliminate the source of sepsis. However, there are conditions when the treatment of underlying disease is impossible (post massive haemotransfusions, aortocoronary bypass, etc.), then we are left with only ALI/ARDS supportive therapy aimed at cupping inflammatory processes and maintaining adequate oxygenation to the tissues.

The progression of parenchymatous lung injury may be induced by excessive ventilation with high tidal volumes, and without PEEP. This will cause cytokine-induced inflammation, now known as ventilation-induced lung injury.

The logical initial goal of ARDS is to achieve PO2 >60 mmHg (corresponding with Sats >90%), as this the upper point of an oxygen dissociation, where the curve bends. Below this level the saturation falls rapidly. The airway pressure must then be corrected to achieve the lowest FiO2. The oxygenation will be stable at least 12 h, high pressure in airways should be reduced gradually, as weaning PEEP rapidly may cause deterioration of alveolar ventilation. Therefore, the PEEP needs to be dropped no more than by '2' every 6 h.

There is no single recommendation to ventilation in ARDS which can bring to a mortality drop. However, the use low tidal volumes (4-6 ml/kg) and high PEEP will maintain the protective lung's ventilation modules. Another recommendation: patients proning and NO inhalation therapy. These improve oxygenation more than other methods.

Restoration of the vitalism reserve is facilitated by ultrafiltration of blood and extracorporeal oxygenation.

5.4. The Role of Enteral Oxygenation (EO)

Enteral oxygenation (EO) in vitalism restoration has not yet reached objective assessment. In respect of the special place the intestine plays in the genesis of multi-organ failure, the question of adequate oxygenation of splanchnic zone's organs is crucial. Therapy aimed at the reduction and elimination of bowel ischaemia is a priority, since the disruption in of a barrier leads to increased insemination of bacteria of intestinal flora, change in motor function, lack of absorption of nutrition and development of inflammatory processes. There is a close correlation between oxygen transport and liver function impairment. Increasing oxygen delivery leads to a reduction of hepatic dysfunction and reduces mortality in critical patients. The search for new methods to eliminate the energy-structural deficit in the intestinal walls has led to the possibility of the development and implementation of early enteral feeding in some categories of patient, which reduces the stress damage to intestinal mucosa. In addition, early feeding tackles the frequency of complications, shortens duration of ventilation and thus the patient's stay in ITU. Early nutrition also helps to increase the albumin level and the activity of phagocytosis, fights bacterial translocation, normalises of glucagon, cytokines, lactate and protein C levels.

The study of Stocker et al. demonstrated that energy-substrate provision of enterocytes depends on the intake of oxygen and nutrition from the intestinal lumen is 50–80%. Therefore, it is in experimental conditions and practice that the use of oxygenated solutions is justified, as it normalises pH of the intestinal mucosa and restores its structural integrity. It is acknowledged that the liver circulation is carried out by the hepatic artery on 35%, and by portal vein on 65–70 %. Therefore, it becomes obvious that oxygen supply to the intestine and a liver can be modified by using enteral oxygenation.

Application of gastroenteric oxygenation is the complex treatment of various diseases which has been ongoing for decades. Originally it was tried on patients with a liver failure only. Based on severity hepatic ischaemic changes to hepatocytes, it is reasonable to perform arterialisation of blood which enters the liver. Increasing the oxygen delivery to arteries is performed by improving the pumping properties of a heart. Vena porta oxygenation is carried out by two ways: either by bypassing the artery and portal vein, or by infusing a certain amount of oxygen into the intestine, assuming that it will be absorbed and transferred to a liver together with the nutrients.

The EO method was proposed by Speransky in 1923. In 1940 he also established that the rate in which the oxygen absorption happens in the small intestine was 0.15 ml/cm sq./h, and in the large intestine: 0.11 ml/sq.cm/h. In terms of area covered by small and large intestine, providing the circulation of blood in mesenteric vessels is intact, it is about 300ml/min, or a half of optimum required oxygen capacity.

Revision of the importance of enteral nutrition in critical patients opens the new door for the opportunity for intestinal oxygen insufflation.

The use of oxygenated amino acid solutions, as well as the intestinal use of oxygenated perftoran (synthetic IV solution), leads to reduction of splanchnic ischaemia and early resolution of gastrointestinal failure. In addition, it has already proven by Gelman (1975) that the increased saturation of venous blood with oxygen through a dosed intestinal injection can assume a beneficial effect of EO in the degree of hypoxic venous vasoconstriction, responsible for manifestation of nosogenous energy-structural deficit in oxygen supply system.

Practically enteral oxygenation is done by the dosed injection of oxygen into the intestine intermittently at specific time intervals, without causing an increase in the intra-abdominal pressure.

5.5. Microcirculation Component of Status-Correction

It includes following pharmacological groups of drugs:

NITROGENS: The use of nitrates to correct microcirculatory disorders is not effective enough, and in some cases may even exacerbate the symptoms of the ischaemia. This can be explained by the facts of nitrates:

- they do not dilate vessels less than 300 mkm in diameter

- they do not change rheological blood properties, and in the presence of hyper viscosity, nitrate-induced shunting can lead to increased arterio-venous difference O2
- prolonged nitrate usage suppresses the production NO in the endothelium and causes tachyphylaxis, disturbs the functional endothelium state and increases platelet aggregation (not significant for large vessels but damaging for capillaries)
- excessive NO accumulation, together with microvascular ischaemia, may potentiate further development of inflammatory reaction, activation of free-radical oxidation and more blood flow disruption in microcirculation

B-Blockers:

- without vasodilating effects that do not affect blood flow in vessels of microcirculation in patients with relatively preserved vessels. However, in the opposite situation b-blockers may exacerbate microcirculatory ischaemia due to their negative inotropic effect, which compromises a tissue perfusion at the level of the microcirculatory net. Also B-adrenoblockers, except for some (carvedilol, nebivolol), practically do not change any pathological mechanism of microvascular ischaemia (which an endothelium suffers from) as well as leave intact blood rheology.

Calcium Channels Blockers

- reduce the functional activity of blood cells and improve functional state of capillary endothelium
- all above without affecting the blood viscosity and red cells deformability

Ace Inhibitors:

- could stimulate and release of endothelin NO
- have antithrombotic and anti-inflammatory properties

Antiaggregants:

- Given the important role of the lists of intravascular thrombosis in the microcirculation blockage, the use of these drugs is mandatory in all patients with a vitalism deficit, as well as patients with microvascular ischaemia
- it has an endothelium-protective effect, together with an antiaggregant one

For correction of all types of microcirculatory disorder, a combination of anti-aggregants with endothelial protectors is most effective

In addition to those mentioned above, respectively, in hemorheological failure and hypercoagulation syndrome, methods are used that can actively perform unblocking of microcirculation and initiate the de-plasmination of interstitial space.

In principle they can be divided into two groups.

EFFERENT: Active interstitium-draining plasmapheresis (AIDP), and microcirculation drainage (MD)

AFFERENT; thrombolytic therapy (TT), heparin-activated cryo transfusion with anti-enzymes protection.

The purpose of the application of AIDP is the retrograde transfusion of interstitial intracellular transudate to intravascular sector. This method is one of oldest adaptation mechanisms of blood loss and part of adaptation syndrome. If this performed correctly the rates of endolymphatic drainage increases significantly.

Most likely during AIDP simulated phenomenon of natural haemodilution, initially there is intersectoral transition from intact interstitium, followed by a response of damaged/ischaemic tissue. This tissue is saturated with plasma-infiltration of myocardia-depressant (MD), ischaemic, necrotic and other pathological substances.

MD allows to exclude the important pathogenic factor: reduction of blood vessel capacity (vasoconstriction), which occurs in response to bleeding. This is achieved due to a fact that the blood withdrawal rate at the stage of haemodilution is correlated with the intensity of retrograde transition of interstitial fluid to the intravascular space.

Another important condition for a drainage of the microcirculation is the presence of adequate blood volume, which provides vasomotion and by optimising the venous return allows the heart to work in the energy-saving mode.

The ongoing de-plasmination of interstitium is evidenced by an increase in concentration of total proteins, with no albumin transfused.

Thanks to AIDP and MD, a decrease in mortality in nosogenous disorders of energy-structural interaction is noted.

Heparin-activated cryo-plasma transfusion with anti-enzymes protection is carried out in order to increase AT-III content and prevent proteolytic damage of tissue due to excessive enzymes load. Cryoplasm contains at least twice as much AT-III as plasma has.

It is the main inhibitor of thrombin, using its covalent bonding, X-A, XI-A factors as well. He also inactivated factors: XI and XII A Neutralisation rate of serine protease of AT-III in the heparin absence is small, with its presence it increased by 100,000 times.

Another property of heparin (so AT-III) is inhibition of plasmin and kallikrein. It diminishes fibrinolysis, controlling both the clotting and anticoagulation system.

Although each afferent method individually has proven to be effective in draining microcirculation, and anti-thrombotic therapy has become the 'gold standard', it is better to start from afferent thrombolytic therapy. And then the microcirculation cleared, it helps to use interstitium –draining plasmapheresis for de-plasmatisation, following heparin-activated cryo-plasma to take control of coagulation.

One of the biggest dangers in energy-structural destruction is iatrogenic hypothermia. Preventing this complication is much easier than treating it. Warming up fluids and using warming blankets and mattresses is a way of dealing with this problem. The effect of severe hypothermia (T< 35 degrees) include reduced drugs metabolism, reversible platelets dysfunction and coagulation impairment. Progressive hypothermia also leads to low cardiac index and dysrhythmia. Compensatory thermogenesis (shivering) contributes to degree of metabolic acidosis and deterioration in a patient's condition.

5.6 Status Resuscitation in Patho-Energy Biothy

Patho-energy biothy is caused by hypoxic ischaemic alteration of BCM, which if spreads lead to a dangerous decrease in energy-productive and energy-protective activity. Its progression means that the number of functional structures of an organism is depleted so much that it becomes unable to provide its energy consumption at least level of functioning, necessary to activate body protective mechanisms (Table 5.2)

Characteristics of Vitalism	Base-line of hypo-energy, %	Status correction, %	Status resuscitation, %
Deficit of vitalism	37 ÷ 19,4	36,1	20,7
Myocardial reserve	38,8 ÷ 48,8	37,8	42,2
Oxygen transport impairment	46,6 ÷ 21,8	46,5	23,4
Micro-circulatory-mitochondrial insufficiency	46,5 ÷ 21,9	44,4	22,9
Hyper-osmolar destability	3,1 ÷ 1,6	2,8	1,3
Diastolic destability	25,8 ÷ 18,1	25,1	18,1
Destructivity	22,3 ÷ 0,58	21,6	2,16
Liability	46,5 ÷ 21,8	45,1	23,3
Inadequacy	68,8 ÷ 22,3	66,7	25,9

Table 5.2 Correction of hypo-energy impairment BCM

The reason for such changes is a deficiency of two energy-protective reserves: oxygen and micro circular and mitochondrial one. An insufficiency of each one reaches almost half the energy-structural demand. The preservation of myocardial reserve provides high anti-hypoxic and anti-ischaemic status correction ability. Autoregulation of energy-osmolar and hemodynamic activity is a destabilising but not life-threatening condition. Nevertheless, the energy-structural relationship and its autoregulation in patho-biothy does not match the energy requirement mainly due to its instability as well as destructivity.

Adding energy-cellular resuscitation to status-correction limits the destructivity of energy-structural status and halves the exhaustion of oxygen transport and microcirculatory-mitochondrial reserve. Diastolic destabilisation hemodynamic energy-protectivity is preserved, witnessing the need to continue the combined status-resuscitation, in order to increase systolic AP by the value of diastolic destabilisation. The hemodynamic component of status-resuscitation becomes energy-protective with AP 85/55 and HR 60–65 per min

The status resuscitation should be carried out continuously, in full range, until a reserve of vitalism occurs. This is characterised by sustainable stability.

This is an evidence of the cessation of perinozal destruction of BCM and a signal to provide energy substrates with a potential of 760–800 kcal/m2/day

5.7. Cell Mass Catabolic Destruction Status-Resuscitation

Hypermetabolism results in excess of ATP, on which consumes more than a quarter of delivered O2, at around 140 ml/min/sq.m. The same amount O2 is used by a healthy adult, whose body is adapted to energy-structural variations. Therefore, a status correction is worth starting from reducing the severity of hyperoxia and the normalisation of oxygen transport and microcirculatory-mitochondrial reserves. (Tabl 5.3)

Characteristics of Vitalism	Base-line of catabolic, %	Status correction, %	Status resuscitation, %
Deficit of vitalism	39 ÷ 58,7	41,6	19,8
Myocardial reserve	20 ÷ 3,8	15	56,5
Oxygen transport reserve	27 ÷ 25	28	16
Micro-circulatory-mitochondrial reserve	27 ÷ 25	28	15,7
Hypo-osmolar destability	8,65 ÷ 13,2	6,3	3,15
Systolic destability	÷28,4	20,5	6
Adaptivity	139,1 ÷ 158,7	141,5	119,8
Liability	27,1 ÷ 25,2	28,1	15,8
Adequacy	112 ÷ 132,8	113,4	104

Table 5.3 Correction of catabolic impairment BCM

Therefore, the usual status-correcting standard, including in catabolism controlled normothermia, and Ca-channel blockers in order to protect vitalism, should be complemented by energy-cellular resuscitation. Autoregulation of factors, responsible for haemodynamics, characterised by systolic destabilisation, requires an inevitable reduction of BP to 165/105 if HR is 105–110 per minute. Stability of destructivity BMC in catabolic patho-energy biothy is caused by 9–13% of hypo-osmolar impairment of energy-structural interaction, which itself deforming and deters the integrity of mitochondria and cells alone. Insensible water losses should be supported with diuretics. Indicative reduction of catabolism intensity should be accounted not just for antioxidants, controlled normothermia but also dehydration, enable to reduce the surface of mitochondria, which decrease the natural 'protons pores ', and declining production ATP.

Hypo-osmolar destabilisation increases free water content in the lungs and a brain, which diminishes the need for diuretics. Due to status resuscitation, the risk of BCM destruction is almost halved, as well as a degree reduction of energy-protective reserves. BCM will then regain adaptivity and reach a normal osmotic state. All of this will indicate the end of catabolic state, and its transition to hyperergy. The latter is characterised by patho-energybiotic insufficiency of energy-structural damage, which can only be corrected by energy-cellular resuscitation.

CHAPTER 6

Remodeling of Vitalism

6.1 Generic Disintegration of Energy-Structural Coupling

Multiorgan failure (MOF) is the universal disintegrating lesion of all organs and tissues, caused by aggressive mediators of the critical condition, with intermittent prevalence of symptoms of one or another organ insufficiency: lungs, heart, kidney, etc.

This definition of MOF is not generally accepted and is more often seen as a simultaneous or sequential damage to vital body symptoms.

In MOF development, one cannot underestimate the disintegrating brain damage, which is manifested by the loss of ability to provide adaptability and stability of energy-structural body equilibrium.

The foundation for the preservation of the overall integrity of a brain is the autoregulation of cerebral blood flow – the ability of brain vessels to change its diameter according to cerebral perfusion pressure (CPP).

Therefore, when CPP increases, the brain vessels contract, the intracranial vascular resistance increases and cerebral blood flow volume decreases, dropping by this intracranial pressure (ICP). Next, with decreasing ICP, the brain vessel diameter increases, and cerebral vascular resistance drops causing an increase in intracranial blood volume and ICP. It should be considered that it takes 30–180 seconds to omit the CPP change and to establish a new level of cerebral vascular resistance.

Therefore. maintaining an adequate level of haemodynamics allows to minimise sharp variation in CPP with subsequent fluctuations of cerebral blood flow and ICP. A decrease in CPP leads to compensatory cerebral vessel dilatation, which allows raise cerebral blood flow and ICP, which in turn, further reduces CPP, forming a vicious circle.

The functional assessment of MOF is shown in Table 6.1

Organ or a system	Baseline function	Insufficiency/failure
Lungs	Hypoxia, needed a ventilation	PCRV, PEEP 10, FiO$_2$ > 50%
Cardiovascular system	Drop in CI, leaky vessels	Inotrops, vasopressors, coronary perfusion maintenance
Coagulation failure	Platelets < 80	DIC
Liver	Drop in a function at least half	Encephalopathy, coma
Kidneys	Oliguria (UO < 50 ml/h), high creatinine	Lactic, metabolic acidosis
Intestine	Paralitic ileus	Obstruction, ischemic bowel, bleeding, Idiopathic cholecystitis
CNS	Confusion, disorientation	Pre-coma, coma
Immune system	Symptoms of inflammatory disease/ sepsis	Immuno-supression, neutropenic sepsis

Table 6.1 Functional MOF assessment

CNS injury manifests by a loss in mechanisms of neural inhibition (due to focal ischaemia) and/or interruption of structural connections (due to concussion) according to the size of the brain damage. A pathological circle is created that determines the appearance of neuropathic syndromes based on neurochemical and molecular processes disturbances.

The most frequent cerebral presentations of secondary nosogenous disease are cerebral oedema, hypotension, raised ICP, vascular spasm, cognitive disorders, delirium, stroke (ischaemic or haemorrhagic).

In contrast to interstitial form of cerebral oedema, 'osmotic leakage' leads to a clinically significant increase in intracellular fluid volume. As a rule, this process will end up in decrease of cerebral blood flow, haematoencephalic barrier breakage and the development of vasogenic oedema.

There are several forms of intracellular cerebral oedema.

ISCHAEMIC OEDEMA: Unlike vasogenic, this is primarily formed not in white matter but in the cortex. The main reason for its occurrence is the lack of sodium-potassium pump, due to a deficiency of energy. In case of primary ischaemic oedema, or reperfusion, the liquid moves from the capillaries to both extra- and intracellular space (post changes).

CYTOTOXIC OEDEMA: Defined out of ischaemic and post ischaemic states when neuronal dysfunction is the consequence of exposure to various poisons, viral infection, intoxication and other similar pathogens.

Neuronal alteration in dys-osmia and dys-hydria causes endogenous metabolic, toxic and iatrogenic imbalances of sodium, potassium, glucose, calcium and magnesium ions between the extra – and intracellular compartments, as well as dissociated osmolarity disorders between these sectors.

The most sensitive cells to the action of dys-osmotic and hydro-ions alteration are those that perform transport functions (vascular endothelium, astrocytes, ependymocytes of vascular plexuses of a brain), as well as cell generating electrical potentials (CNS neurons and cardiomyocytes.) An acute increase in osmotic blood pressure leads to opening of inter-endothelial contact, and to a sharp increase in CNS permeability microvasculature, following perivascular haemorrhages

Sharp iatrogenic osmolarity changes extracellularly, combined with critical cell energy deficiency, modifies hydrophilic channels of the plasmalemma and impairs an active transport of water and ions neurons and glial cells. Rapid water shifts increase osmotic pressure, which leads to acute hyperhydration (osmotic swelling) of cells. The consequences of dys-osmotic and hypotonic alteration are acute disorders of ionic and acid-base balance between neurons, glyocytes and extracellular space of CNS, which change process polarisation of nerve cells' membranes dramatically and can result in a coma.

6.2 Splanchnic Ischaemia

This is caused by intra-abdominal hypertension syndrome (IAHS), which is an increase in intra-abdominal pressure up to 20 mmHg or more and is accompanied by organ dysfunction and even failure.

Shortness of breath is the first manifestation of IAHS. The biomechanics of breathing (axillary muscles involvement, an increase in the oxygen cost of breathing) suffers significantly. With its progression, acute respiratory failure develops rapidly, and a patient needs the respiratory support quickly.

Kidney function impairment at the beginning IAHS does not reflect CI dropping, but is the consequence of abdominal compartment syndrome, which also lead to increasing resistance of the kidneys vessels, decreased kidney perfusion (drop in renal blood flow) and then glomerular filtration rate. It is all apparent as intra-abdominal pressure (IAP) reaches 10–15 mmHg, and anuria is seen at IAP.30 mmHg.

Intra-abdominal compression leads to the disturbance of microcirculation and thrombosis in small vessels, ischaemia of the intestinal wall, following translocation of bacteria and its toxins into mesenteric bloodstream and lymph nodes.

When IAP reaches 30 mmHg, the lymph flow along with thoracic duct is disrupted and seized.

With the portal blood flow decreasing and IAP >20 mmHg, the metabolism, including drugs, decreases. Studies shown when IAP 25 mmHg reaches and above it can cause decrease in a brain perfusion pressure and results in cranial hypertension.

At present, surgical decompression (laparotomy) is the only treatment of IAHS. The normal IAP considered 12 mmHg.

The clinical equivalent of developing gut ischaemia is intestinal insufficiency syndrome (abdominal obstruction). It is characterised by impaired digestive, transport and barrier functions of the intestine. That is why the intestine (with its source of toxins) is considered the main reason for the development of multiorgan failure.

Abdominal obstruction is seen in early stages of IAHS. Changes in intestinal wall permeability in late stages may cause the systemic infection, sepsis and MOF in relation to endotoxin and bacterial translocation. Disturbances in normal micro-equilibrium in the intestine are caused by excessive colonisation/growth of certain bacteria, especially enterococcus, which leads to bacterial translocation. The probability of it is higher in patients in critical conditions, with long antibiotics used orally, with no antibacterial drugs cover.

Many (if not all) protective mechanisms which should prevent bacteria spread in patients with multiorgan failure are suppressed. These patients are often immunocompromised, and multiple courses of antibiotics in their treatment contribute to disappearance normal intestinal microflora, giving a green light to colonisation of pathogenic microorganisms

The most pronounced microcirculation disorders occur in the proximal gut (stomach, duodenum) due to highest content of alpha-adrenoreceptors. This is reflected in the structural and functional changes of these organs. Typical motor-evacuation disorders are: gastroduodenal dyskinesia, lack of pyloric sphincter and duodenogastric reflux. These play an important role in the pathogenesis of erosive and ulcerative gastrointestinal lesions.

6.3 Acute Kidney Injury

Acute kidney injury (AKI) – a syndrome occurring of acute decrease in tangle filtration, that often develops in critical patients with multiorgan failure.

The new adopted classification of AKI is based on the level of severity of its injury. The classification's name is RIFLE, formed from: Risk, Injury, Failure, Loss, End.

As the latent form of AKI is masked by the clinical picture of the underlying condition, early diagnosis is only possible to distinguish by investigating diuresis dynamic systematically, as well as checking urine markers. Only at insufficiency stage (as per RIFLE) clinical symptoms may occur, caused by direct kidney function shutdown. In case of late diagnosis, or at loss of control of water balance, there are signs of hypervolemia (petit oedema. or swelling of subcutaneous fatty tissue), congestive heart failure (small blood circle). Severe cases manifest by pulmonary oedema.

Hyperkalaemia is a very frequent and severe complication of the oliguric AKI. If not treated, hyperkalaemia can cause a sudden cardiac arrest, the danger of which increases sharply with present metabolic acidosis.

Hyponatremia (sodium plasma < 135 mmol/l) is another dangerous complication of AKI.

Moderate hyponatremia is mostly asymptomatic, but sometimes can be accompanied by gastrointestinal disorders. CNS disorders are detected when Na of plasma is <125 mmol/l. It is mainly due to cerebral oedema. Increasing cerebral oedema leads to a coma.

According to a triggering pathogenic mechanism, pre-renal, renal and post-renal AKI are distinguished. Various causes can lead to different forms of AKI develops.

Between 45% and 74% (60% in average) of critical patients will require renal replacement therapy in AKI.

An ultimate evidence of criticality of abdominal ischaemia is ACUTE LIVER FAILURE. Histological presentation of this as such:

- hepatocellular necrosis with or without preserved liver architectonics
- lobular collapse
- islets of hepatocytes regeneration
- infiltration by polymorphonuclear cells, lymphocytes, plasmacytes and eosinophils with a proliferation of structures around portal zones.

High mortality is associated with development of the following complications:

- cerebral oedema
- renal failure
- sepsis
- pancreatitis
- heart failure
- multiorgan failure (MOF)

6.4 Uncontrolled Proteolysis

Proteolytic enzymes attract special attention of clinicians by their participation in the development of shock, malignancy, and many pathological conditions accompanied by the destruction of energy-structural interaction. Normally, there is a balance between production of proteolytic ferments and their inhibition, which can be disturbed in case of general or local body damage. As the rule it causes the activation of proteolysis. This, in turn, leads to initiation of humoral and cellular protection systems, such as inflammation, stimulation of antibodies production as well as cellular immunomodulators, do not disregard an activation of plasma proteolytic circles: haemo-coagulation cascade, fibrinolysis, kallikrein-kinin, renin-angiotensin and system of complement.

Proteinases take control for organ and tissue functions, causing the formation, modification and initiation of virtually and biologically active substances, including enzymes, regulatory proteins and peptides.

There is a reason to believe that uncontrolled proteolysis in biological destruction can distort the function of proteolytic plasma systems. This will disrupt processing and degradation of peptides that regulate intestinal motorics. It is currently known around 30 peptides are involved in the regulation of bowel movements, including vasoactive intestinal peptide, cholecystokinin, substance P, opiate and others forms, which are processed by specific proteins of gastrointestinal and APUD cells under hyper-proteolysis conditions, due to the release of powerful lysosomal ferments, such as elastase, cathepsin G, thiol, matrix, etc. In addition, these proteins damage receptors of regulatory peptides, hormones, biogenic amines, etc. Hyper-proteolysis receptors and peptides, controlling intestinal motility, may be the one of the reasons for ileus in biological destruction.

One of destructive proteinases: elastase, plays an important role. It inactivates pre-kallikrein, kallikrein and Hageman's factor, which violates the control of all proteolytic systems of blood. To stabilise the Hageman factor, it is expedient to use 20–30,000 of units of kontrical (proteolytic drug), or aprotinin, which possess the same properties.

Fibronectin (FN) – is high-molecular glycoprotein, which regulates the intercellular connections, such as phagocytosis, inflammation, regeneration and haemostasis. There reactions are carried out due to the presence of FN sites (domains), specific to cell receptors, and components of intercellular matrix, like collagen, heparin, hyaluronic acid, fibrin, fibrinogen, as well as bacteria and bacterial toxins. Studies have been obtained to show the therapeutic effect of FN FFP to enhance phagocytic function of RES and granulocytes.

FN induces adhesion of cells in the glycocalyx, and serves as chemotaxis for leukocytes, endotheliocytes. The chemotaxis is done by through reversible and alternating processes of adhesion and breakage, associated with the contractile elements of cytoskeleton, which induces and enhances the bacterial attachment to the phagocyte's membranes with its following immersion to the cytoplasm, with activation of bactericidal processes in phagocytes of RES.

FN is also a regulating factor in vascular permeability, as it is a part of the basal capillary membrane. With its deficiency, for instance in alveolar-capillary membrane, the interstitial lungs oedema can manifest. To some extent this evidence explains direct dependence of arterial PO2 on FN level. The destruction of FN significantly decreases with the introduction of counterbalance – an inhibitor of plasmin, trypsin and other proteolytic ferments. In the hypercatabolic state, deterioration of utilisation of standard energy subtraction, amino-acids imbalance and interruption of protein synthesis, leads to development of protein-energetic insufficiency like hypo-fibronecitinemia. AT-III reduces the activity of proteolytic enzymes and FN destruction, as well as its consumption in the intravascular clot formation, which is an important point in maintaining adequate amounts of this protein.

On the other side, the activation of own proteolytic enzymes of the pancreas may lead to acute pancreatitis (AP). This is a form of acute inflammatory-degenerative disease of the pancreas, based on autolysis of pancreatic tissue by its enzymes, with further development of aseptic or microbial inflammation. The leaking pancreas is a source of damage to surrounding organs and systems of extraperitoneal origin.

The most possible mechanism of intra-acinar activation of trypsinogen and other zymogens is the hypothesis of colocalization (thickening). In early stages of AP, the lysosomal enzymes thicken together with the zymogenic granules, producing trypsin (with the influence of cathepsin B).

Zymogenic activation triggers next circle of events: release of phosphodiesterase A2 to bloodstream (which destroys surfactant): elastase presence initiates haemorrhagic reactions, and a lipase load leads to peritoneal and extraperitoneal fat tissue necrosis.

A violation of kallikrein-kinin system and complementary systems develops, and systemic inflammatory response flourishes resulting in shock and MOF.

It is important to note that all plasma proteins became a target of proteolytic ferments leaking from necrotising pancreatic cells. Therefore, the blood coagulation system and fibrinolysis have a clear face-like character. In this case a therapy aims to regain control over coagulation, increasing anticoagulation potential and timely activation of immuno-genesis.

Following the similar scenario, a persistent lipid metabolism disorder develops, not only in the pancreas but in target organs. The changes in the phospholipid layer of the bio membranes of cells in various organs manifests in the first 24 h of a disease. The crucial modifications of lipid membrane composition affect the main lipid layer, which gives a ground to define this pathological phenomenon as a membrane destabilising process. One of the consequences of the above is impaired lipid metabolism and an excessive formation of phospholipid lysophores, which, like fatty acids, have a damaging effect. Another negative impact of changed lipid metabolism in tissue of various organs is change in lipid composition of blood plasma and lymph, strengthening of free radical reactions of lipid peroxidation, reduction of antioxidant enzymes, and phospholipase system activation.

The frequency of extra organic complications in AP reaches 60–96%. The clinical outcome of severe systemic inflammation is sepsis and a development of MOF.

6.5 Symptomatic Therapy in Vitalism Imbalance

Initially the severity of MOF is determined by the number of organs affected or the systems that are failing. According to one group of physicians, if the number of failing systems increases from 1 to 4 the mortality increases from 30 % to 100 %. Others agree that with three systems affected the patient has no chance to recover.

Staging of actions – it is the main principle of the strategy to combat MOF, which Zilber specifies as:

1. The first stage is to provide artificial support or replacement for a system (or systems), without replacement in which the lethal outcome is inevitable. Most likely examples of these systems are breathing and circulation.
2. This results in a 'pause' providing the opportunity to conduct preliminary multifunctional examination and het to give at least a rough idea of the degree of damage to organs and systems.
3. Efforts should be directed at the correction of pathophysiological mechanisms, affecting all systems, for example, the application of anti-inflammatory therapy correction of metabolism, etc. Such a coordinating influence on all systems allows to improve some of it.
4. Finally, we must deal with the organs that remain effected and help them recover gradually.

Energy biocorrection as a strategic principle of combating bio-disintegration has not yet become dominant, although post syndrome therapy reflects energy protective intentions. The latter may include:

1. Metabolic correction. First, the maintenance of acid-base equilibrium – a lack of which causes the 'working capacity' of enzymes involved in the production of energy to be compromised.
2. Adequate, balanced nutrition, enteral and parenteral.
3. Introduction of vitamins and essential amino acids to normalise enzyme activity.
4. Adequate oxygen delivery to tissues, which sometimes requires lung ventilation and microcirculation drainage.
5. Antioxidant and anti-inflammatory drugs.

A sequence of actions, focussed on functional correction of vital organs and systems, is presented in Fig 6.2

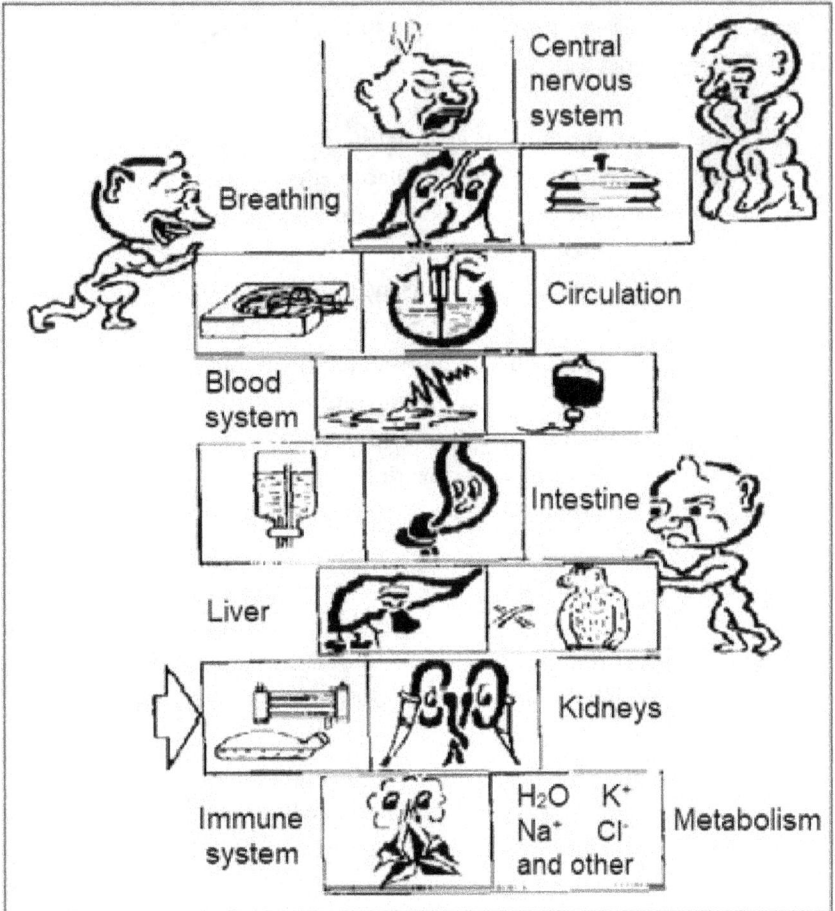

Fig 6.2 Functional correction of vital organs and its functions (by Zilber A.P)

To summarise, MOF syndrome may include hypovolemia, coma, respiratory distress, clotting system impairment, circulatory failure, acute renal or hepatic failure, etc. Syndrome-relevant therapy of each of these conditions is a complex of methods aimed at artificial substitution or support of certain organs and systems, without considering the nature of energy-structural interaction. (Fig 6.3)

Organs	Stage of MOF	Correction
Lungs	Dysfunction	Dispnea, $PaO_2/FiO_2 > 250$ torr, Oxygenation with FM/NS
	Insufficiency	Hypoxia, PaO_2/FiO_2 150-250 torr, NIV, Ventilation
	Decompensation	Ventilation over 72 h, PaO_2/FiO_2 75-150 torr, ECMO
Heart and vessels	Dysfunction	BP stable, PVP normal, Isovolemia, Myocardial protection
	Insufficiency	Hypothention, Microcirculation failure, Inotrops, vasopressors
	Decompensation	Unstable BP, IABP
Kidneys	Dysfunction	Oliguria, Creat < 250 μmol/L, Pre-renal failure: IV fluids, diuretics
	Insufficiency	Hyperazotemia, Creat > 250 μmol/L, Hyperkaliemia, Diuretics
	Decompensation	Oliguria over 36 hours, Potassium > 7 mmol/L, CVVH
Liver	Dysfunction	Hepatocytolisis, Increased liver enzymes, Billirubin < 60 μmol/L, Hepatoprotectors
	Insufficiency	Albumin < 20g/L, PT < 60%, Ф XIII < 50%
	Decompensation	Hepatic coma, Upper GI bleeding, Liver cirrhosis
Intestine	Dysfunction	Enteropathy, malabsorbtion, ileus, Pro-kinetics
	Insufficiency	Bowel obstruction, erosions, Ulcers: bleeding, Laparatomy, drainage
	Decompensation	GI bleed, ischaemic bowel, OGD, laparatomy, bowel resection
Central nervous system	Dysfunction	Confusion, drowsiness, GCS 10-14
	Insufficiency	Coma, GCS 6-9, Intubation, airway protection, CT head, unclear prognosis
	Decompensation	GCS < 6, bad prognosis

Table 6.3 Clinical and biochemical orientation in organ impairment with prioritization of the therapy

Special attention should be paid to a treatment of intestinal insufficiency/ failure. The treatment is a complex of therapeutic measures aimed at the elimination of morpho-functional disorders of the gastrointestinal tract with the transition to early enteral nutrition, which can decrease the reduction of the splanchnic circulation, pathognomy to MOF. The enteral feed should contain W3 fatty acids (3–4 g per day), and balanced proteins and carbohydrates load.

The full decompression of a stomach and small intestine supported by drugs stimulating motility, as well as initiation of intestinal lavage and entero-sorption, or even with hyperbaric oxygenation help.

As has been said, early NG feeding is a method of choice that provides the most natural and adequate way to convert nutrients.

Early stimulation of immunity, in this case intestinal immunomodulation, is an important stage in preventing intestinal bacterial translocation and release of endotoxins into the bloodstream, and then organs and tissues.

To prevent stress ulcers, bolus injections of IV PPI's are used (once or twice, 40 mg).

For a prevention relapse of bleeding, continuous PPI infusions are administered (8mg/h) for 72 h, following 20 mg omeprazole daily.

PPI drugs are preferable as H2- inhibitors are not effective in bleeding

The complex of therapeutic measures in AKI includes correction of hypovolemia, usage inotropes, treatment of hyperkalaemia, hyponatremia, acidosis, or sometimes elimination of hypervolemia as it causes lung oedema.

Particular attention is paid to restoring circulating blood volume to normalise cardiovascular function. In patients with cardiovascular diseases, or in conditions where a fluid is redistributed into third space, recovery of intravascular blood volume should be done under CVP control.

The use of furosemide does not reduce the needs for renal replacement therapy (haemodialysis, hemofiltration). In some cases, such as pulmonary oedema, diuretics are ineffective, CVVH is the treatment choice.

The liver has a remarkable ability to regenerate, that is why in MOF the treatment of liver failure consists mainly of maintenance therapy until the restoration of organ function.

Treatment of encephalopathy, associated with acute liver failure, is aimed at limiting the formation of ammonia in the intestine and preventing aggravating factors like infections, intestinal obstruction, ileus, gastrointestinal bleeding.

For the purpose of suppression of ammonia-producing intestinal flora, the commencing of metronidazole (not in a case of liver failure/encephalopathy as it neurotoxic) or neomycin (knowing about its nephron- and oto-toxic effect) are useful.

Acute liver insufficiency is a catabolic state with rapidly developing disorders of protein and energy metabolism. Therefore, enteral feeding is preferable to parental feeding.

Fever and leucocytosis are absent in 30% of infected patients with acute liver failure. Prophylactic antibacterial therapy protocols for acute hepatic insufficiency do not exist, but the use of broad-spectrum antibiotics and antifungal agents leads to positive effects.

Circulatory hyperdynamy is a characteristic feature of acute liver failure. With systemic vasodilation and dilatation of blood vessels of internal organs it leads to an increase in CO and a decrease in BP. Volume replenishment is necessary for the correction of arterial hypotension, but normal pressure values are rarely achieved. This must be done under control of CVP, to ensure the adequate circulatory volume is reached.

The only radical treatment for decompensated liver failure is liver transplantation.

ACUTE CARDIAC FAILURE (ACF) is a frequent companion of MOF, characterised by rapid onset of symptoms systolic and diastolic insufficiency (low stroke volume, CI, insufficient tissue perfusion, pulmonary hypertension, petit oedema).

Critical reduction in CI – activates neurohumoral mechanism (renin-angiotensin and sympatho-adrenal system), leading to a further increase in systemic vascular resistance and deterioration of myocardial contractility. As a result, oxygen diffusion deteriorates, following a fall in contractility, which aggravates the acute cardiac failure. (Fig 6.4)

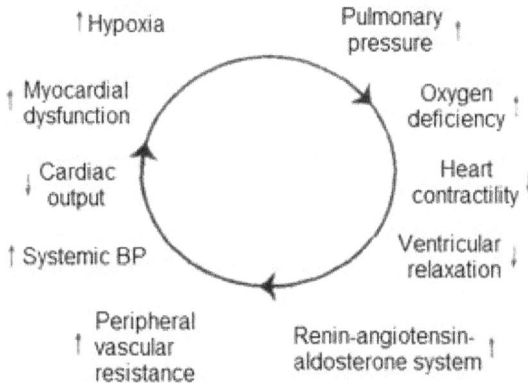

Fig 6.4 Pathogenesis of development of acute cardiac failure

101

The 'vicious circle' can start at any level, but all the changes make the patient worse.

One of the modern methods of diagnostics and outcome prediction in heart failure is determination of brain sodium uretic peptide (BNP) level.

Treatment of ACF should start from maintaining adequate oxygenation (Sats >95%). Non-invasive ventilation with positive pressure should be preferred for respiratory support. Ventilation with intubation is only recommended if respiratory failure is unable to control with support of oxygen therapy, NIV and vasodilators. Apart from in cases of pulmonary oedema in acute coronary syndrome, then ventilation is a treatment of choice.

Diuretics are used in acute coronary syndrome and in chronic cardiac insufficiency.

In the treatment of ACF other drugs are necessary, such as inotropes, to reach a reasonable BP. However, they increase myocardial oxygen demand as well as the risk of rhythm disorders. Even a short-term use of inotropes can cause its adverse outcome. Another indication for inotropes usage is the presence of peripheral hypoperfusion (arterial hypotension and deterioration of kidney function), even with lungs overload, refractory to diuretics and vasodilators in optimal doses.

At present, the doctors are equipped with a new inotropic drug: levosimendan. Unlike other inotropes, levosimendan has two main mechanisms of actions: the sensitisation of contractile cardiomyocytes proteins by calcium ions (positive inotropic effect), and activation of potassium channels, causing peripheral dilatation. Also, levosimendan has a mild inhibiting effect on phosphodiesterase. In addition, levosimendan has an active metabolite with similar properties, so the action of the drugs depends on a concentration of thistalite. The halftime of this active substance is 80 h, that determines the long haemodynamic effect of a drug administration.

Levosimendan is given to patients with ACF, caused by systolic dysfunction without significant hypotension (BP >85 mm. Hg). Levosimendan is given by an infusion 0.05 -0.1 mcg/kg/min (at least 10 min), and then reaching 12-24 mcg/kg. Hemodynamic efficiency of this drug is dose-dependent, and it can be infused with 0.2 mcg/kg/min maximum

6.6 Methods of Detoxication

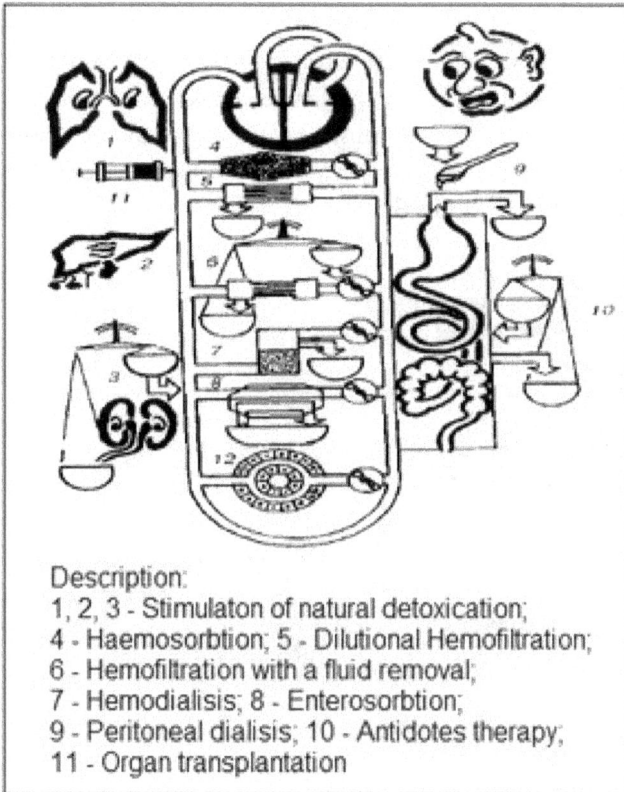

Description:
1, 2, 3 - Stimulaton of natural detoxication;
4 - Haemosorbtion; 5 - Dilutional Hemofiltration;
6 - Hemofiltration with a fluid removal;
7 - Hemodialisis; 8 - Enterosorbtion;
9 - Peritoneal dialisis; 10 - Antidotes therapy;
11 - Organ transplantation

Fig 6.5 Methods of detoxication (by Zilber A.P.)

Detoxication infusion crystalloids and blood products as the action of clearance in the systemic blood flow is considered one of the simplest therapeutic actions in terms of active detoxication. The increase in plasma blood volume and the expansion of extracellular liquid space reduces the extracellular and intravascular concentration of endogenous toxic substances (ETS), contributes to a decrease in the severity of endogenous intoxication (EI) and inhibits the progression of endotoxicosis.

Forced diuresis (FD) is a sequence of infusions, medications and actions, aimed at maintaining high diuresis for a few hours or days. In clinical practice, urine output of more than 2 ml/kg/h, about 3.5 times or more above the normal diuresis, is considered as forced.

A necessary condition of boosting diuresis is the absence of deep necrobiotic changes in the filtration membrane and tubular kidney epithelium, as well as the preservation of adequate (hyperdynamic) reaction of cardiovascular system to excess fluids load, acute hypervolemia and haemodilution.

ENTEROCORBTION – is the active detoxication method, based on binding endogenous toxins and substances, sub molecular structures and cells in the lumen of gastrointestinal tract, as well as its removal for therapeutic and preventative purposes. (Table 6.6)

Direct effects	Side effects
Sorbtion endotoxins came from bile, pancreatic juice, sectets of serous fluids	Decreasing load on functioning body detoxication system
Sorbtion endotoxins of hydrolisis	Correction immunostatus, improvement of endohomeostasis
Sorbtion of biologically active substanses: neuropeptides, prostoglandins, histamine, serotonin, anti-allergens	Correcting of disbalance of biologically active substanses
Sorbtion bacterial toxines, pathogenic bacterial bodies and its parts, microbic elements of intestine	Returning vascular bowel permeability for normal intestinal bio flora, its toxins and metabolites
Sorbtion intestinal gases	Treatment of ileus, bloating, improving vascularisation of an abdomen
Irritation of the receptor zones in the gastrointestinal tract	Stimulation of intestinal motility

Table 6.5 Direct and indirect enterosorbtion effects (by Kostiuchenko, Gurevich, Gulakov)

The idea of this method is the use of enteral sorbents – different preparation of medicinal substances, which carry out the binding properties and extract toxins or other pathogenic elements from gastrointestinal tract by absorption, ion exchange and complexes formation.

The therapeutic effect of entero-sorption is determined by summarising its direct and indirect effects.

DRUG ELUTED INCREASE BIOTRANSFORMATION EFFECTIVITY is not fully understood. Thus, active sodium hydrochloride, after a period of active application, has become limited in use...'

6.7 Circulation Drainage

Depending on coagulation screen numbers, the cryo-plasmatic-anti-enzyme complex for microcirculation drainage is used in three variations (Tseyman)

OPTION 1: Moderate doses of FFP: 300–450 ml + high doses of heparin (2500 IU/every 100 ml of FFP), added to a plasma transfusion, together with 5000 IU 4 times daily subcutaneously

Indications: hypercoagulation, slight decrease in AT-III (75–85%), plasminogen reserve index: 75–90%, thrombocytopenia (ortho-phenanthroline test up to 13.5 g/l x 10 (square) slight increase in F-r

OPTION 2: Large doses FFP (600-800 ml) IV in 2 infusions, small or medium doses of heparin (2500 IU/100 ml of FFP), with 2500 IU x 4 times/day subcutaneously, average doses of protease inhibitors (kontrikal 80000 ATU TDS and 4000ATU- 5000 ATU intravenously in the following 3 days, or Gordox 300000 ATU in 12 h/day with 150000 twice daily in the following 3–5 days.

Indication: Marked decrease in AT-III (65-75%), a decrease in plasminogen activity (65-75%), weakening of XII-dependable fibrinolysis, moderate thrombocytopenia (OFT: 13.5 -18.5 g/l x 10 (square) moderate increase PDF and PMK.

OPTION 3: High doses of FFP (600-800ml), with small heparin doses: 2500 IU/100 ml FFP, high doses protease inhibitors (kontrikal 100000 ATU three times/day in the first day with 500000 IU of it per day next 2–3 days, or Gordox 50000 IU first 24 h, following 300000 IU next 2-3 days.

Indication: hypercoagulation, haemorrhagic syndrome, hyperfibrinolysis, hyperfibrinogenaemia, acute deficiency of AT-III <65%, plasminogen < 65%, marked thrombopenia >18.5 g/l x 10 (square), high increase PDF and PMK.

In addition to well-known methods of drug administration, endolymphatic therapy is used in MOF. The existing methods of direct endolymphatic therapy (antegrade and retrograde) requires surgical intervention, which is a significant disadvantage of a method. Indirect endolymphatic drainage, or regional lymphatic therapy: drugs administration through a surgically inserted catheter into the liver, stomach, intestinal mesentery (Petrov) is very promising. The endolymphatic therapy with lymph stimulation for motility stimulation is bases on the endolymphatic injection of lymphangion-thyrotropin-releasing hormone (TRH- 12.5 mcg/in 20 ml saline) and on increasing lymphatic drainage

One of highly effective methods of extra-corporal detoxication is plasmapheresis.

The blood exchange volume depends on the endo-toxicosis level and symptoms MOF. Usually in 2 cycles of blood circulation volume are removed at about 30% of circulating plasma volume (the rate of exchange is 1 ml/kg/min on average).

Heparin as a preservative is given in a dose 7–8 units/ml of blood in vitro. The volume of plasma-replacing solutions corresponds with a volume of removed plasma as 1:1.

There are two methods of centrifuge plasmapheresis used:

1. Intermittent: discrete with the use of plastic containers as 'HEMACON': 500/300, with refrigerator centrifuge RC -6 in 2000 circles/min at 22 degree for 15 min
2. Continuous: method with using a blood separator (see a picture)

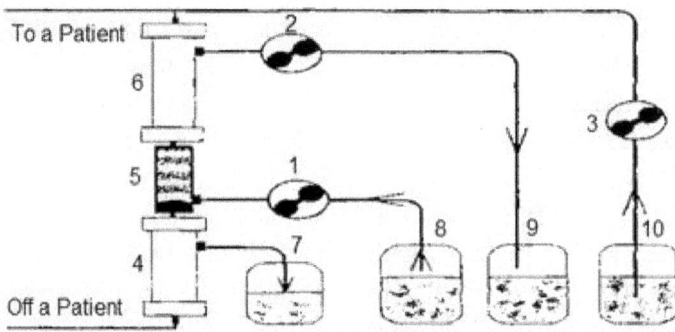

Fig 6.7 Scheme of plasmoferesis

Description:
1,2,3 – roller pumps
4 – plasma filter
5 – column of washed erythrocytes
6 – haemofilter
7 – container for collecting plasma
8 – solution for erythrocytes wash
9 – container for washing liquid collection
10 – plasma-substituting solution

An essential part of plasmapheresis exchange is the issue of plasma replacement and the prevention of possible protein deficiency. In protein plasma substitutions above 60–65%: the protein deficiency is not on evidence.

In 2–3 % of body mass plasma removal, with 60–65% protein transfused there is no change in haematocrit (Ht). This itself indicates the adequacy of the plasma exchange volume with the preservation of fluid balance.

If plasmapheresis is performed in patients with a tendency to hypotension, you may experience: tachycardia, hypotension, though to correct these, the plasmapheresis should be stopped, and restarted only after hypovolemia correction.

In plasma filtration, even in unstable patients, during a process, and even after it, the tendency to hypotension is non-existent.

Dr. Vatazin developed a method selective extracorporeal detoxication (SED), where blood is taken from portal vein. Then it used bi-filtration cascade exchange plasmapheresis, optimal hemofiltration and plasma sorption.

During the procedure, apart from plasma exchange, the washing and liquid oxygenation of erythrocytes is used in a special designed device. Toxins, washed from the cells' membranes, together with washing up solution are filtered through a haemofilter. The volume of one session of such plasmapheresis reaches about 800,3 +/- 60 ml. Plasma loss is compensated by an adequate number of donors FFP and albumin. The volume of ultrafiltration contains 24.6 +/- 1.2 l. The volume of perfused plasma reaches 1840+/- 20 ml.

The detoxication function of a liver, which became an obstacle to toxins penetration through a portal vein system to systemic blood stream, is restored in 3 days.

The application of low-flow blood membrane oxygenation in liver failure ensures the increase in partial oxygen pressure in a blood, and CO_2 exchange. Application of the latter method in patients with MOF is proven to be effective due to multifunctional impact on the patient's body.

6.8. Main Pathways of Vitalism Remodelling

These are determined by energy-biometric peculiarities of biological disintegration manifestation.

Both types are characterised by the pre- or terminal decrease in energy delivery due to dis-hydric hypobiothy, which is responsible for the mismatch between over-increasing ultracellular compartment and progressive reduction in the amplitude genome's expression.

Therefore, the clinical reflection of the severity of suppression of signal transduction from becoming an inactive receptor to body cell mass genetic apparatus, is the resistance or refractivity to the measures correlating nosogenic arterial hypotension.

Vitalism prosthesis is based on infusions and inotropic support, and it's difficult to implement as it inevitably leads to intracellular hyperhydration and acidosis. Consequently, the correction should be strictly dose dependent: the volume of infusion is limited CVP, and amount of sympathomimetics are used to reach the oxygen-protective level of diastolic blood pressure (DBP). The latter number should provide adequate coronary perfusion pressure and minimise the myocardium afterload.

Practically adequate circulation volumes are maintained by '5-2' rule, following an adrenaline use to increase systolic pressure to 70 mmHg. Once it is reached, it is possible to switch to milrinone (dopamine, dobutamine) inotropic support, the next goal will be stabilising a diastolic pressure (up to 60 mmHg). The minimal crystalloids load and reduced inotropic support can be achieved by using lympho-drainage 'suit'. The absence of hemodynamic effect is due to desensitisation of adrenoreceptors.

This condition can be treated with hydrocortisone and insulin. Resistance or ever more hemodynamic refractivity usually caused by blood shunting in the lungs (up to 30% of CI), pulmonary hypertension (up to 60 mm Hg) and hypoxic venous, which reduces coronary and systemic blood flow19 stronger than arterial spasm

Disintegration	Parameters
Resistent stage	Resistant hypotention; Reversable patho-energoprotection with pre-terminal energy-structural status; Stressed energy-conjjugation; Dishydric insufficiency of energy-structural status
Refractory stage	Refractory arterial hypotention; Resistant patho-energy protection; Refractory patho-energybiothy with terminal energy-structural status; Dishydric failure energy-structural status

Tabl.6.8 Characteriscics of disintegrated biological activity

Reduction in pulmonary shunting and arterial hypoxemia only available with artificial ventilation, leading to increase in intra-alveolar pressure, decrease in venous return afterload with diverting blood from lung circulation to systemic one.

Ventilation will help with eliminating second dead space (with inverted cycles phases), provide oxygenation (FOI2) and normocapnia (PCO2). To switch to ventilation, use hypnotic: propofol, barbiturates: midazolam, relaxants. Good relaxation is required to cap the stress, which can manifest with increased vasospasm. The characteristics of preferable ventilation: small tidal volume (6–7 ml/g), PEEP, if required high frequency ventilation. Compared to standard APV technique, low TV does not cause additional formation of pro-inflammatory cytokines, as later are a trigger factor in the pathogenesis of MOF. Minimisation of FIO2, to keep Sats 92–94% allows to control inflammation.

6.9 Dyshydric Nature of Disintegration

The lack of an acceptable solution to the problem of bio-disintegration underlines the consistently high level of mortality in MOF. In patients with MOF the direct cause of death is not a disease itself (nothing is left to indicate a disease on post-mortem), but deep changes in the organs' parenchyma, caused by energy-structural coupling disorder. In other words, the complex treatment program allows elimination of the main pathological factor and thus to increase the life expectancy of patients.

However, it does not prevent the development of progressing several pathological conditions of energy-structural interaction, which predetermined the tragic outcome. These pathological conditions manifested by signs of systemic or organ failure, becoming independent, taking a lead from first symptoms recurrence, and despite the reduction of nosogenous pathology cause the lethal changes. The clinical picture of MOF was quite clearly delineated and included respiratory distress syndrome, hepatic renal failure, formation of stress ulcers, coagulopathy, and damage to the integrative ability of CNS.

This situation allows us to conclude that MOF is not only clinical or pathophysiological in character, but also a problem in vitalology in no lesser degree. It is no coincidence that interest in the biological nature of MOF continues, and many studies have raised this question, in most general terms, but not in specific form.

All proposed theories, largely developed based on different directions (ethology, pathogenesis, clinical picture), or on a basis of some symptoms of pathological process, are far from solving the nodal issue of a problem. The efficacy of MOF treatment turned out to be much lower than theoretically assumed, which gives a rise to the negative attitude of clinicians to these perceptions, which are nowadays, to some extent, losing their significance.

The failure of these theories is due to a fact that MOF has been conceived as a purely pathological process devoid of any protective functions. At the same time, nature has invented the principles of adaptive reduction or even termination of functions. In the state of abiosis, the organism retains a real possibility of life resumption on condition that the corresponding functions will be restored.

The generalised stage of the energy-structural peri-nosological development of vitalism deficiency is presented in the Fig.6.9

ENERGYBIOTHY	HYPER	HYPO / PATHO	NOZOGENOUS ETIOLOGY	HYPER / NORMO / HYPO	PATHO	ENERGY PROTECTIVITY
	HYPER		**ACTIVATION** OF ENERGY-STRUCTURAL INTERACTION maintenance eubiothy preservation of vitalism	HYPER		ENERGY
			REALISATION OF ENERGY-STRUCTURAL INTERACTION decreasing reserve of vitalism cutting down of energy consumption artuficial eubiothy	NORMO		
	HYPO		**ENERGY-STRUCTURAL IMPAIRMENT** deficit of vitalism energy-cellular resuscitation resuscitation of vitalism	HYPO		PROTECTIVITY
		PATHO	**ENERGY-STRUCTURAL DAMAGE** life threatening deficit of vitalism status -cellular resuscitation controlled vitalism declining		PATHO	
			DESINTEGRATION OF ENERGY-STRUCTURAL INTERACTION non-reversable deficit of vitalism protesis of histo-hematic balance tendency to dropping vitalism			
			BIOLOGICAL COMPLEXITY OF A BODY			

Fig.6.9 Stagers of deficit vitalism and its correction

The main protective components of the hypobiotic reaction are reduction in oxygen demand, water and energy components, as well as sensitivity to hypoxia. Also, the increase of tolerance to toxins, preservation energy and functional resources necessary to exiting the hypobiotic reaction, following prevention of functional and organic damage (Sherman). However, such an adaptation significantly restricts the body's ability to fight for a life, in comparison with its capacity of active reaction.

The trigger mechanism of bio-disintegration is nosogenous activation, which can cause an increase in BCM cytosol osmolality by 40–50 mosmol/l on average. Inevitably, by the law of iso-osmolality, intracellular hyperhydration is necessary to equalise osmotic pressure.

Hypothesis testing of the water role in development of deficit of vitalism in MOF is done in experiments with rats. It was found that the development MOF was accompanied by an increase in total water, a sharp decrease in a free fraction, and an increase in content of bounded water.

The obtained results allow us to judge not only about a degree and sequence of organ lesions in MOF, but also provide an opportunity to evaluate the effectiveness of drug therapy in MOF. The mathematical regularities obtained during the results analysis, provide a base for further development of options of intensive care in MOF.

6.10 Prosthesis of Histo-Hematic Exchange

Modelling of histo-hematic metabolism should be started after achieving good oxygenation, normocarbia and ensuring of SVP and MAP at the oxygen-protective level. Currently, hemofiltration (HF) is a method of choice for MOF with RF in oliguric phase.

HF is used for elimination of inflammatory mediators like cytokines, complementary complexes, peroxide lipids, free oxygen radicals and bacterial toxins. Other anti-shock effects of HF include restoration of iso-osmolarity, as well as nitrogen oxide synthesis in the splanchnic region of gut, increasing venous return to a heart, elimination of cardio-depressive factor and proteins, causing apoptosis and death of parenchymatous cells.

In addition to main effects, the possibility of delivering large amounts of energy (lactate, glucose) carried by HF deserves attention. Another important impact of HF is the elimination of dyshydria and marked improvement in the rheological properties of a blood under the influence of poly-ionic substituent. The value of timing and intensity of HF in the beginning of prosthesis for reversal of intracellular decompartmentalization, organ dysfunction and prevention MOF irreversibility, has been established.

The leading indication for HF is the need for correction for severe metabolic, cardio-respiratory acidosis, prominent intoxication, electrolytes disbalance and encephalopathy. After ultrafiltration, blood is returned to the venous system oxygenated, decreasing pulmonary vasoconstriction. For HF used standard compensated electrolytes solutions according to insensible losses.

Modelling hydro-ion-osmotic properties of BCM may increase vitalism and reduce mortality in MOF, presented with renal failure. This could be the method of biological correction if parameters of HF are chosen wisely.

At the beginning of a process, it could be a drop in haemodynamics, if the circulating volume is depleted.

Significantly suffers a body oxygen capacity. Deterioration of the lungs' gas exchange accompanies severe arterial hypoxia if the circulating volume does not match the required one. In combination with hemodynamic changes, deterioration in gas exchange leads to a reduction in oxygen delivery. The pulmonary shunt, which can be controlled by intra-aortic pumping of some blood volume from the venous circuit, is responsible for its reduction. Even if both hemoperfusion safety conditions are not met, there is a tendency to restore hemodynamic equilibrium after HF. In the nearest postperfusion period, the haemodynamics tends to recover.

Carrying out hemofiltration in patients with MOF gives both positive and negative effects, with the latter being more frequent as the vitalism control is not maintained. The very process of HF is accompanied by drop in BP and oxygen supply until oxygen-protective iso-osmia is reached. In the absence of stage-by-stage osmo correction, which ensures the elimination of damaged body cells, hypoxia remains responsible for the patient's survival. Close relationship of CI, oxygen transport and oxygen consumption indicate the only possible way to keep an adequate oxygenation it must be a hyperdynamic type of circulation. However, this is not always possible in kidney failure, as increasing volume of infusions (to maintain preload), and the use of inotropes, in the absence of energy-biotic monitoring can lead to disruption of the compensatory capacity of circulating volume.

Further prospects of hardcore enrgy-biothy correction are opened at joint use of plasmafiltration, plasma sorption and hemofiltration.

Practically justified to filter and le-ligate plasma, that which is discarded post filtration then returns to the patient. The outcome of bio-prosthesis is primarily determined by the rate of elimination of hypobiothy by restoring the vitalism deficit, which highly depends on the extent of expression of hypoxic genes.

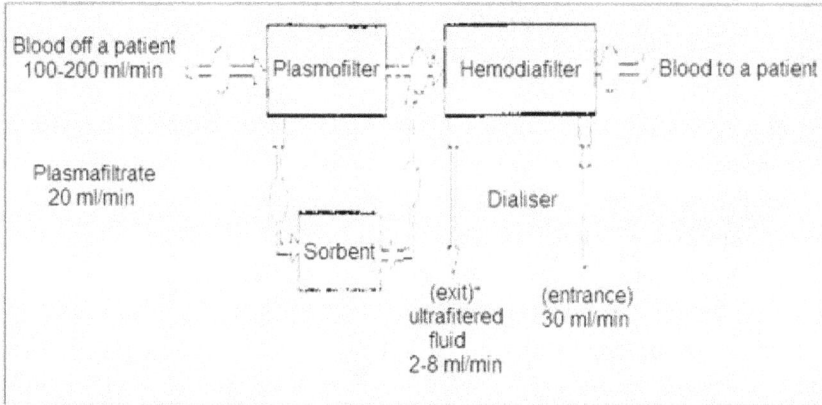

Fig.6.10 Schemes of plasma filtration, plasma sorption, hemofiltration and remodeling of histohematic exchange

The duration of vitalism-prosthesis is determined by a time of recovery adequate oxygen delivery and acid-base state normalisation. The need for a repeated procedure is maintained as long as systemic intravascular alteration is present.

6.11 Energy Efficiency of Vitalism Remodeling (Prosthesis)

Prosthesis of multiorgan energy-structural insolvency is lifesaving (Fig. 6.7)

Characteristics of vitalism	Base-line of multi-organ failure, %	Prosthetics, %
Deficit of vitalism	80÷66	63
Myocardial reserve	÷33	32
Oxygen transport insufficiency	99÷87	77
Micro-circulatory-mitochondrial insufficiency	63÷47	44
Hyper-osmolar destabilization	4,3÷2,8	2,5
Dyastolic destabilization	÷9,8	4,5
Systolic destabilization	4,7÷	÷
Destructivity	33÷22	20
Lability	62,6÷46,6	43,7
Inadequacy	95,6÷68,6	63,7

Fig.6.11 Deficit, reserve , auto-regulation and vitalism properties in MOF

113

As it can be seen from Fig 6.7, the risk of MOF is irrevocable due to reserves exhaustion and development of severe oxygen deficiency. The life is maintained by myocardial reserve of energy protection and at least 50% of microcirculation and mitochondrial apparatus preserved. The precise autoregulation remained haemodynamics is beneficial for the survival in conditions of arterial hypotension. The resulting hyperosmolar BCM destabilisation is tolerable and should not become an object of correction for achieving water-electrolyte balance due to a danger of hypo-osmolar deformation of cyto-architectonics.

Inadequacy of energy supply to a body needs can become absolute. Therefore, the early onset and continuity of complex extracorporeal vitalism-prosthetic technologies determines the dynamics of restoring life forces. Initial favourable shifts are expressed by a decrease the danger of BCM destruction, a fall in oxygen deficiency, drop in destructivity and liability of the vitalism

After remodelling, MOF is replaced by pathobiotic instability, successful overcoming which requires consistent, stage-relevant restoration of vitalism, the use of status-resuscitation, and then energy-cellular resuscitation, which serve as the main purpose of vitalism prosthesis.

CHAPTER 7

Medical Practice of Programmed Information System 'Vitalism Audit'

7.1 Abstract, Primary and Refinement Data

The effectiveness of medical treatment will increase with audit of vitalism in medical care. It applies to the understanding of the requirement for simultaneous treatment of diseases and patients alone. This goal will allow the information system (IS), 'Vitalism Audit' (VA), based on the idea that vitalism is a manifestation of the life forces of the BMC.

Comparison of the current energy production with the level of its reliability and maximum capacity allows one to establish the borders of vitalism, its reserves and adaptability.

The information system 'Vitalism Audit' is able to give an objective assessment of severity the patient condition, to determine the safety of used complex of perinosal medicine, to establish the danger of applied treatments and recommend energy- cellular technologies to eliminate the vitalism deficiency

Primary Obligatory Parameters
Date ... Time
Name ... Hospital number
Diagnosis
Comorbidities
Component of treatment:
Weight, kg ... Height, cm
Age
SBP (systolic BP)
DAP (diastolic BP)
HR (heart rate/min)

Refinement Data

Hb, g/l

SaO2, % oxygen saturation of a blood

a/v difference O2 (20-90%)

MBV (minute blood volume), ml/min

SV (stroke volume, ml)

Cx – optimal a/v O2, %

Qx – factor of compensation in oxygen delivery, % (0,4 – 1.8)

Blood osmolality (mosmol/l) (250-340)

Na concentration, mmol/l (125-170)

K concentration, mosm/l (1.8 -9.0)

Glucose in blood, mmol/l (0.8 -20)

Urea, mmol/l (1–30)

Temperature, C (33–43)

7.2 Parameters of Audit

DV – Deficit vitalism, % (1–90)

RV – Reserve vitalism, % (1–50)

VO2 – Oxygen consumption ml/min/sq.m (30-550) dependable of level of activity

RVO2 – required oxygen consumption ml/min/sq.m (90–150)

AVO2 – actual (not measured) oxygen consumption (70–250)

CI – cardiac index, ml/min/sq.m (1–8)

BSA – Body's surface area, sq.m (1–2.6)

MBV – Minute blood volume, ml (1.8–15)

LBV – Left ventricle volume, ml (20–150)

SBP – Systolic BP, mmHg

DBP – Diastolic BP

MAP, mmHg

Age, years (14–120)

Temperature

Weight, kg

HR – heart rate

RHR – required heart rate

RMBV – required minute blood volume, ml

RCI – required cardiac index

BMR – base metabolic rate, kkal/24h (210-4400)

RBMR – required base metabolic rate, kkal/24h (900 -1800)

MER – myocardial energo-structural reserve, % (1 -300)

OESR – Oxygen energo-structural reserve, % (10-60)

AL – Activity level, ml/min/sq.m (>110 ml/min)

RL – Reliability level (readiness to go), kkal/24h (900-1800)

P – Level the maximum energy production, ml/min/sq.m (70-250)

A – Area of energy-structural stability, ml/min/sq.m (130-160)

Energobyota, % EB

EB 2: 85–147% EB- hypoenergybyota Changes: 1 – 65

EB 3: 148–192% EB- euenergybyota 101 – 125

EB 4: 193 and >% EB – hyperenergybyota 1 – 320

EB 1: 83 and < % EB – pathoenergybyota 1 – 70

OTR – oxygen transport required (rage 60 – 2000)

ED – energy delivery (100 (VO2/OD), %

VO2 – 1.34 x Hb x SaO2 x CI, (actual), ml/min/sq.m

OD – oxygen delivery (estimated, calculated)

1,34 x Hb (male/female standard) x 0.96 x CI(estimated), ml/min/sq.m

Rage: 220–900

Hb, estimated (male) 132 g/l

Hb, estimated (female) 120 g/l

Energydynamy, %

ED1: 84% and< % ED – pathoenergyprotectivity Changes: 1–70

ED2: 85 – 147 % ED – hypoenergyprotectivity 1–65

ED3: 148 -192 % ED – normoenergyprotectivity 101–127

ED4: 193 and > % ED – hyperenergyprotectivity 1–320

Biological Stattus (BS, ml/l)

BSES Energystructural Stability 160–130

BSHD Hypoergic Disfunction 129–105

BSHyperD Hyperergyc Disfunction 161–185

BSD Hypoergic Damage 104–85

BSHyperD Hyperergic Damage 186–210

BSHI Hypoergic Insufficiency 84–75

BSHyperI Hyperergic Insufficiency 211 – >
BSEF Energystructural Failure 74 – <

Properties
SBP (systolic blood pressure) (designed BP> average BP) – adaptivity, %
Range: 90–250
DBP (Diastolic blood pressure) (Average BP>designed BP) – destructivity, % 1–60
SS (Actual BP > designed BP) – stability stage, % 100–150
IS (designed BP > actual BP) – Instability stage, % 1–75
AS – adequate stage, % 101–180
IAS – inadequate stage, % 2–135
MMESR – Microcirculary-mitochondrial energy-structural reserve, % 10–80
MESF – Myocardial energy-structural failure, % 1–50
TESD – Transport energy-structural deficiency, % 1–50
MMESD – Microcirculatory- mitochondrial energy0structural deficiency, % 1–50
O2TESD- Oxygen transport energy-structural deficiency, % 1–50
Qx – Compensatory factor energy-structural needs for O2, 0.4–1.8
Sa O2 – oxygen saturation of arterial blood, % 40–96
SvO2 – oxygen saturation of venous blood, % 30–92

Energybioty, %
OOS – Oxygen-osmolar stability, % 100–170
SOP – Plasma osmolality required for stabilisation
energy-structural status, mosm/l 270–330
RPO – Required plasma osmolality, mosm/l 280–310
HOD – Hyperosmolar destabilisation energy-structural activity, % 1–20
HypoOD- Hypoosmolar destabilisation of ES activity, % 1–20
HDS – Hemodynamic stability of ES activity, 0.599–0.636
DD – Diastolic destabilisation of ES activity, % 10–25
SD – Systolic destabilisation of ES activity, % 5–25
COP – Calculated plasma osmolality of ES activity, % 270–310

7.3 Audit of Perinosal Vitalism

OCC = (148 – VVO2)/1.48, %
OCC – oxygen capacity coefficient,

With VVO2 >/= 192, OCC = 100 x (VVO2 -193/VVO2, %

A – Level of in energy-structural activity equal VVO2

S – Standard ES level (average) male/female figures

D – designed O2 consumption of ES complex, VVO2

R – Level of reabilility of ES complex, if calculated cx, then VVO2 = Cx x CI

If measured SvO2: VO2 = CI x 1,34 x Hb x (SaO -SvO2)

If unable to measure SvO2: VO2 = CI +0.382)/o.026, ml/min x m.square

CI = MBV/BS, ml/min x sq.m, where MBV – minute blood volume

BS – body surface area

BS = V height(cm) x weight (kg)/3600

MBV = SV x HR

If SV unable to measure: 0.5 x (SBP +DBP) – 0.6 DBP – 0.6 x age, ml

Average MBV = K x (100 + 1.33 if male/1.44, if female)

If male age >61y.o or female >56 y.o, it's not 1.33 but 1.5

DMBV – designed MBV = K x (100...

DDBP – designed diastolic BP = 63 + 0.4 x age, mmHg

DSBP – designed systolic BP = DDBP/0.618, mmHg

DHR – designed heart rate = 48 x (hight, cm/weight, kg)/3

DCI – designed cardiac index = DMBV/BS (body surface area)

PA – present ES activity

PA = VO2 x 7.07, kcal/day

PAC = ES consumption

PAC = VVO2 x 7.07

MPAC = ES consumptions in males

MPAC = 66.47 + 13.75 x weight + 5 x height – 6.78 x age

FPAC = ES consumption in females

FPAC = 65.5 +9.56 x weight +1.85 x height – 4.67 x age

7.4 Energystructural Reserves (ESR)

CESR – Cardiac reserve = (220-age>HR)

CESRK – cardiac reserve coefficient = (220-age-HR)/HR, %

CESD – myocardial ES deficiency (HR> 220-age)

CESDC – CESRD coefficient = (HR -220 + age)/HR, %

CTESR – oxygen delivery ESR: Q< 1(Q -shunt)

CTESRC – CTESR coefficient = 100 x (1-Q), %

CTESD – ES oxygen deficiency, Q>1
CTESDC – CTESD coefficient = 100 x (Q-1), %
Q = VVO2/VO2, subjected units
MMESR- microcircular-mitochondrial ES reserve, %, av >Cx
MMESRC – MMESR coefficient= 100 x (av-Cx)/av, %
MMESD – MM ES deficiency, %
MMESDC – MMESD coefficient = 100 x (Cx-av)/Cx, %
Cx = VVO2/CI, if av measured
If av not measured, but: SvO2 available: av = 1.34 x Hb x (saO2 -SvO2), ml/min
or av = VO2/CI, ml/min
SVO2 = SVO2 m/f/7.07

Properties ESA
VOR – vitalism oxygen reserve (VVO2 range 149 -193)
VORC – VOR coefficient, (VVO2 -148)/VO2 x 100, %
A – adaptivity, % (Designed > Average), %
D – destructivity (Average> Designed), %
S – stability, (Actual > designed), %
SC – stability coefficient= 100 +100x (VO2 – VVO2)/VVO2
DF – deficiency (liability), (Average> Actual), %
DFC – deficiency coefficient = 100 x (VVO2 0VO2)/VVO2, %
AD – adequacy, %
ADC – AD coefficient = AD + DF
NAD- non-adequacy, %
NADC – non-adequacy coefficient = D + NAD, %

7.5 Autoregulation of Energyprotectivity

ABS – acid-base stabilisation
ABSC – ABS coefficient= (DESPO – APO) x DESPO/APO, %, where
DESPO – designed ES plasma osmolality
APO – actual plasma osmolality
0.97 x APO < DESPO < 1.03 x APO
HOD – hyperosmolar destabilisation ES activity, %
HODC – hyperosmolar destabilisation coefficient= 100 x (APO -DESPO)/
APO, %

HODS – Hypoosmolar destabilisation ES activity, %
HODSC – Hypoosmolar destabilisation coefficient ES activity
100 x (ABS –ABSC)/ABSD, %
HDS – hemodynamic stability ES activity
HDSC – hemodynamic stability coefficient in ES stability
100 x ((DBP/SBP)/ (0.599 – 0.636)) + 100
HDDS – hemodynamic diastolic destabilisation of SE activity,
(For DBP/SBP <0.636)
HDDSC – hemodynamic diastolic destabilisation coefficient
100 x (DBP/SBP – 0.636)/(DBP/SBP), %
SDS – systolic destabilisation of ES activity
(For DBP/SBP < 0.599)
SDSC – systolic destabilisation coefficient
100 x (0.599 –DBP /SBP) / (DBP/SBP), %

7.6 Energobioty and Energyprotectivity

EBT – energybioty, %
EBTC – energybioty coefficient = 100 x PA (present ES activity) / designed PA
EBT 1 – 84% of EBT and < pathoenergybyothy, %
EBT 2 – 84- 147 % EBT – hypoenergybiothy , EBT 2 = 147-EBT, %
EBT 3 – 148 – 192 % euenergybiothy, %
EBT 4 – 192% and> – hyperenergybioty, EBT = EBT – 193. %
PT – VO2 x Q, ml/min /m square
ED – energydynamy, %
EDC – ED coefficient = 100 x VO2 / DVO2 (DVO2: designed oxygen transfer)
VO2 = 134 x Hb x 0.96 x CI, ml/min x m square
DVO2 = 1.34 x Hb x 0.96 x DCI (DCI – designed CI) ml/min x m square
Hb (male) = 132 g/l....Hb (female) = 120g/l
ED 1 :84% ED and < – Pathoenergo-protectivity
ED1 = 84 – ED, %
ED 2: 85 –147% – ED Hypoenergo-protectivity
ED2 = 147 –ED, %
ED 3: 148 –192% – Euenergy-protectivity
ED 4: 193 % and > hyper energy-protectivity
ED4 = ED –193. %

OSMP = 332 − 0.026 x VO2 − 0.137 Po2, mOsm/l
DSMP = 332 −0.026 DVO" −0.137 DVo2, mOsm/l
OSMP = 1.86 x (cations + urea + glucose =5), mosm/l
DOSMP = 332 −0.026 VO" −1.13 VVo2 , mosm/where
OSMP – plasma osmolality
DOSMP – designed plasma osmolality

7.7 Resume

Conditions ongoing vitaloresusscitation
BSE - Biol. stability
Standard level BP, HR, osmolality (266-296), metabolism (1040-1160)
BSHD Bio-hypordynamy: BP 120/75 -110/65, HR 80-70/min, Osmolality 301 - 304, Met 995-900kcal
BSHyperD Hyperdynamy: BP10/85 - 150/95, HR 90-100, Osmolality 294-297mOsm
Met 1130-1230kcal
BSP Pathodynamy   BP 105/65 -95/60, HR 70-63, Osmolality 303 -307 mOsm   Met 890 -805kcal
BSPHyp Pathohyper   BP 150/95 - 165/100, HR 110-100, Osmolality 293-290 Met 1237 -1336
BSPHypo Pathohypo   BP 95/57 - 90/55, HR 62-59, Osmolality 307 – 308 mOsm
Met 800-765 kcal
BSHyperH   BP 165/105, not more, HR up to 111, Osmolality 291mOs
Met 1337and <
BSEDeficiency: BP up to 88/54, HR 58 and <, Osmolality 309, Met not less than 760

On a verge of perinosal medicine deficit of vitalism (DV) is ...... (Number A)
On the next measurement   deficit of vitalism   is .... (Number B)
Number "B" is different from "a" by ....   (Number C)
If "B" is smaller than "A" the deficit of vitalism improves, the treatment is working,
if opposite:

the treatment is failing

If "B" = 0: the deficit of vitalism diminished.

If deficit of vitalism non-existent, look for a reserve of vitalism

Next to calculate a reserve of vitalism (RV) is .... (Number A)

On the next measurement the reserve of vitalism is .... (Number B)

Number "B" is different from "a" by .... (Number C)

If "B" is bigger than "A" the deficit of vitalism improves, the treatment is working,

if opposite:

the treatment is failing

If RV is "0" this means a risk of developing a deficit of vital damage....

Assessment of Energy_structural Reserves of Vitalism

Myocardial reserve is ....(Number)

Oxygen transport is ....(Number)

Microcirculatory and mitochondrial reserve i s.........(Number)

These calculations will show if ES level is at risk of ES destruction.

If it is RV (reserve of vitalism): will it be going into developing critical dysfunction

ES activity (ED 2 + 85 - 147%, ED 3 148-192%)

If it is happening the treatment needs to be questioned.

Developing RV (reserve of vitalism) without going in hyperdynamy leading to a good prognosis

Assessment of Properties of ES Vitalism

Energy-structural activity characterised by:

        PEB – pathoenergy bioty

        HEB – hypoenery   bioty

        EEB – euenergy   bioty

HyperEB – Hyperenergy bioty

The one divided to – PEP 1 – pathoenergy-protectivity

        EP 2 – hypoenergy-protectivity

        EP 3 – euenergy- protectivity

        EP4 – hyperenergy- protectivity

In this the activity can vary from A – adaptivity to "D" – destructivity
For example: If A – adaptivity reached   (Actual figure > Average), if where is the
D – destructivity   (Average   > Actual),
as well as (Actual > Designed) – it's a stable state,
with (Actual< Designed) show inadequacy of energy-structural interaction

Reliability of Vitalism
Reliability provided (%) by energy- osmolar and /or hemodynamic stabilisation. Insufficiency ES interaction caused hemodynamic and energy hype (or hypo/hyper osmolar destabilisation.
Haemodynamics may be presented as either systolic or diastolic destabilisation of circulation

Recommendation
EEB + EP: to maintain it is necessary to keep own body auto-regulation
HEB + EP3
EEB + EP2   to correct a dysfunction it is necessary to use artificial eubioty
HyperEB + EP4
HEB + EP2:   to correct insufficiency necessary to use energy-resuscitation
PEB + EP2
HEB + PEB: these 2 conditions treated using status correction
PEB + PEP: to save a life and re-new ESS necessary to start the ESS (energy-structural status) prosthesis

Remodeling of Auto_regulation of Vitalism
Energyosmolar interaction regained by approaching level plasma osmolality (POSM) to designed plasma iso-osmolality (DPOSM
Eyenergyosmolar equilibrium occur when: POSM is about (1+/- 0.02) x DPOSM
In POSM > DPOSM osmo energy protectivity provided by optimizing IV fluids and inotropic support systemic circulation, plus adding cardio tropic. drugs, normalizing urine output.

All these measures reducing hyperosmia.

In hypoosmolality osmoenergy protection maintains infusion of crystalloids. Energo-hemodynamic equilibrium guarantees that the speed of deoxygenation of oxygen is corresponding with normotension level (105 /60 - 140 /90).

or DBP/SBP = 0.5999-0.636. In diastolic destabilisation   DBP (designed BP) = SBP x (1 + HOD), and in systolic: DBP = SBP x (1- SDS) mmHg.

Renewal Vitalism Reserve

To restore the myocardial reserve, we need to drop the heart rate, using drugs.

(HR X (220-age). Oxygen transport deficiency will be eliminated if CI will increase on CI x OTD, or HB will be topped up on 25 x CI x OTD (oxygen transfer deficiency number).

Micro-circulatory and mitochondrial deficiency tackled by controlled normo tension, if a difference on O2 reached (av x (1 + MMD number), ml/l.

Conclusion

The main biological property of the human body, responsible for its integrity, survival and maintaining the special laws of the organisation, is only can be identified at the level of entire organism. And it is called vitalism.(vital, eng - vitality). This organic factor of a life force is determined by interaction energy and structural body capacity, responsible for either healthy or sick status of a patient.

Vitalism alone can become scarce or built in reserve dependably on borders of energy delivery figures which should be matching energy production level. Energy needed for the genetic apparatus of changing BCM for continuous preservation of biological integrity of the organism.

Therefore an auto-regulation has several features. Thus, auto stabilization activity energy-structural status (ESS) BCM requires constancy of cyto-architectonics, which in turn is supported by cytoplasmic reticulum and intracellular fluid.

The ions balance determines osmotic pressure, responsible for transmembrane movement electrolytes and water molecules, which if very, can be a cause of deformity BCM and damage to ESS mechanisms. A specific study of the relationship between the parameters of oxygen transport and osmolality made it possible to present it numerically, in order to be used for provision of energy-osmolar stabilisation. It is preserved if the real osmotic pressure does not deviate more than +/- 3% from a value energy needed for biothy maintained and energy protection.

If the response to energy-osmolar destabilisation takes milliseconds, the hemodynamic response will go up to 2 minutes if not more. This is why even minimal changers in the amplitude of blood pressure, either systolic or diastolic pressure ratio, proved to be crucial. The value of ratios reflects the interaction of forces of total blood flow and its organ distribution, responsible for delivery of oxygen/subtracts to organ/tissues. The ratio equal (0.599-0.636) corresponds with stabilisation of hemodynamic autoregulation of ESF. The increase beyond the upper limit of the interval shows diastolic destabilisation, the decrease beyond the lower limit - systolic. The visibility of the obtained result emphasizes its representation in percentage terms.

Energy structural status reflects its properties: adaptivity or destructiveness, stability or lability, adequacy or inadequacy,

Adequacy testifies to the ability of BCM to provide additional energy-structural support for genetic programs for ESS level of activity.

The destructivity of EES occurs if its activity is unable to increase the energy production or energy-structural support of hypoxic genetic programs responsible for the functioning of BCM at the stress response level.

The ability to increase ESS in response to BCM activation reveals the comparison of the current intensity with the level of optimal energy efficiency. If the latter is lower than the actual one, the inability of BCM to increase its energy efficiency will be maintained.

The lability of EES is evidenced by the failure to fully satisfy the energy requirement of BCM.

The adequacy of ESS to BCM in the implementation of genetic material comprises sum adaptability with lability, while for destructivity and stability is responsible an inadequacy.

Depending on the peculiarities of personal energy-biotic characteristics and energy-protective clinical status of patients, 8 categories of energy-structural statuses were identified.

Determination of the categories and variations of violations responsible for their formation made it possible to develop specific methods of directed therapy capable of ensuring reliability of energy protection and energy preservation.

Euenergybiothy is the key to maintaining discrete energy-protective status by meeting the body basic physiological needs in water, electrolytes and energy substrates

Energy resuscitation is a method of increasing stability of ESS at nosological damage by restoring effective functioning of microcirculation-mitochondrial complexes. It includes correction of dyshydria with iso-osmia, achievement of blood tension volume, maintenance of normocarbia and oxygen-protective values of blood pressure and hematocrit.

Status Correction - a technique for elimination of nosogenous destruction of energy-structural status by restoring acid-base equilibrium and microcirculatory-mitochondrial distress. Thus responsible for an adequate intensity of ventilation and components of oxygen transport, the unblocking microcirculation and terminating plasma leakage.

Status Prosthesis - it is energybio-resuscitation complex measures aimed og regain the stability of biological integrity of an organism using extracorporeal elimination of destructive and decentralising dyshydric damages, and prosthesis of metabolic and microcirculatory processes..

The main factor in the vitalism achieved is played by anti-energy wasting osmolar optimization. Its essence is to bring the current plasma osmolality to the level of minimal energy cost. Then the energy-cellular osmo deficit/excess is determined and the total amount of mOsm/l is calculated to optimize dysosmia.

Stabilisation of energy-structural chaos at loss of vitalism is accelerated if the values of quantum biocycle (QBC) is transformed into the achievement of HR,
determined by Kerdo equation.

Oxygen-hemodynamic harmonisation has a vitalotropic effect on the ESS in that it is able to exclude hypoxic and hypertoxic damage of BCM. It achieved by the ability to synchronize the rate of erythrocyte deoxygenation with the level of normotension in space (105/60 - 140/90 ml Hg)

The relationship of energy-structural interaction will become inherent to the vasomotion if the rhythm of its function corresponds to the rate changers in HCO3, which depends on the intensity of tissue metabolism.

The knowledge provided by vitalology ensures that the treatment process, full of innovative technologies of BIO-protection, is focused on restoring the vitalism reserve.

Possession of the scientific evidences of vitalology and the information system "Vitalism Audit" will ensure accelerated recovery of patients and will increase life expectancy toits biological limit

Bibliography

1. Gayton A.C , Hall J.E Medical physiology - Moscow: Logosphere, 2008-1296 p-------- - А. С. Гайтон Дж. Э. Холл Медицинская физиология , Москва Логосфера 2008- 1296 стр

2. Grippy M.A., - Lung physiology - Moscow : Binom , 2005 - 304 p----------------------------- М. А. Гриппи Физиология легких, Москва : Бином 2005 - 304 стр

3. Zilber A.P , Etudes of respiratory medicine . Moscow: MEDpress -Inform 2006 - 568p -----А. П. Зильбер Этюды респираторной медицины Москва МЕДпрессаИнформ 2006 -568 стр

4 Zilber A.P , Etudes of respiratory medicine . Moscow: MEDpress -Inform 2007 - 792p ----- А. П. Зильбер Этюды респираторной медицины Москва МЕДпрессаИнформ 2007 -792 стр

5.Kolesnik M.A. Tumansky V.A.., Shifrin G.A Principles of medical Competence . Zaporishye: Wild Field, 2013 - 374 p ----Колесник Ю М.,Туманский В А Г А . Шифрин. Принципы врачебной компетентности Запорожье : Дикое поле , 2013 - 374стр

6 Selie G , Essays on Adaptation syndrome - Moscow: Medgiz, 1960 - 254 p ---------------------Г. Селье .Очерки об адаптационном синдроме. Москва
Медгиз ,1960 - 254 стр

7. Usenko L.V. Shifrin G.A. Intensive therapy for blood loss - 3rd edition, conceptual and innovative - Dnepropertrovsk: New ideology , 2007 - 290 p----------
Усенко Л В Шифрин Г А Интенсивная терапия при кровопотере 3-е издание - Днепропетровск Новая Идеология 2007 -290 стр

8. Shifrin G.A. Restoration of bioproperty in sepsis - K. Expert , 2004 - 604 p---------------- Шифрин Г А Восстановление био устойчивости при серсисе
К Эксперт 2004 - 604 стр

9. Shifrin G.A , Kolesnik M.A. Tumansky V.A Vitalologia - Zaporoshye, Wild

Field, 2018, 288 p------------ Шифрин Г A Kolesnik Ю М В .А. Туманский, Виталология Запорожье : Дикое поле , 2018 - 288стр

10. Siggard. O, Andersen L.H. Jothem P.D, Wimberley N., Fogh-Andersen . The oxygen status of arterial blood revised : relevant oxygen parameters for monitoring the arterial oxygen availability. Scand.J Clin.Lab.Investigation, 1990, 50 Suppl 203, 17-28---

11. Shifrin G.A. Homeostasis. Securing therapy and statusmetria . Acta Anestesiology Scand 1995 :39 Suppl .107, 257-259p---------------------------

12. Shoemaker W.C., Appel P.I, Kram H.H . Prospective trial of supranormal values of survivors as therapeutic goals in high-risk surgical patients. Chest.1988
94, 1176 -1186

* 9 7 8 1 9 1 3 3 4 0 4 6 9 *